STATISTICS FOR EVERY FAN

Sports Statistics
SPRINGFIELD COLLEGE

Carl Fetteroll
Springfield College, Massachusetts

Copyright © 2017 Carl Fetteroll

All rights reserved.

ISBN-13: 978-19-759-1844-6 (Gray)
ISBN-10: 19-759-1844-4 (Gray)

Contents

History of Sport Statistics class at Springfield College

PART 1 Descriptive Statistics — 1

1	**Brady. Brees. Manning.** **Who is the best quarterback?**	3
	Quarterback rating exercises	6
2	**Turning numbers into data.** **Is it magic?**	9
3	**Variable variables**	17
	Variables exercise	20
4	**A picture is worth a 1000 words**	21
	Categorical variables exercise	28
5	**Time out for testing**	29
6	**Fit fest stats style**	33
	Chi-squared goodness of fit test example	36
	Chi-squared goodness of fit test exercises	42
7	**Does average make sense?**	45
8	**A picture is worth a 1000 words (take two)**	49
9	**SHAPE, ceNTer & spread**	53
	Shape, center, spread exercise	57
10	**A box with whiskers?**	59
11	**Back-to-back boxplots**	65
	Back-to-back boxplot examples	67
12	**Measuring spread**	69
	Standard deviation example	76
13	**Who had the better day? Z-scores**	79
	Z-score examples	83
	Z-score exercise	87

PART 2 Predictive Statistics — 105

14 Simpson's paradox — 107
Simpson's paradox exercise — 110

15 Title IX: Thou shalt not discriminate — 111
Title IX example — 114
Title IX exercises — 116

16 Trending now in sports — 119
Time-series plot exercise — 122

17 Gaze into your crystal ball — 123

18 Which is the best predictor? — 131

19 Borda count — 135

PART 3 Statistical Inference — 149

20 Confidence intervals for 1 proportion — 151
Proportion exercise — 155

21 Hypothesis testing — 157

22 1-proportion questions — 161
1-proportion example — 167
1-proportion exercises — 169

23 Significance tests about hypothesis — 173
One Sample t test examples — 176
One Sample t test exercise — 179

24 2-sample means — 181
2 sample mean example — 185
2 sample mean exercises — 188

25 Estimates of a population — 193
Proportion exercises — 194

26 2-proportion questions — 197
2-proportion examples — 199
2-proportion exercises — 202
2 proportion and 2 sample means exercises — 205

27 ANOVA — 207
ANOVA example — 209

Appendix 1: Chi-Squared Distribution Table — 233

Statistical techniques are tools of thought, not substitutes for thought.
Abraham Kaplan

Wrigley Field, Chicago, November 3, 2016, the day after the Cubs clinched the World Series Trophy.

Carl Fetteroll
Author

Danny Fetteroll
Illustrator

Carl Fetteroll has been an adjunct math professor at Springfield College in Massachusetts for 16 years. Carl also runs a recreation program for children with visual impairments at Springfield College and is a past coordinator of the Massachusetts Senior Games.

In his spare time he participates in triathlons, completing two Iron Man distance triathlons, the Great Floridian in 2014 and Ironman Quebec in 2015.

Danny Fetteroll is an illustrator and blogger for the Oakland Athletics branch of SB Nation. Born in Boston he has had an odd obsession with the Oakland Athletics since he was a child. To pay the bills he works as a barista at Starbucks which allows him the flexibility to attend plenty of Cubs games.

Danny's Athletics comics and player profiles can be found at sbnation.com/users/fatrolf/blog.

History of Sport Statistics class at Springfield College

In 2005, the math department at Springfield College discussed how to make our elementary statistics course more accessible for our students. Most were taking Algebra assuming it would be easier than stats, since they had all taken it in high school. Plus statistics was thought of as indecipherable and with the huge $200 textbooks filled with problems that students had no interest in, impenetrable.

Looking at our undergraduates, 92% had played sports before coming to Springfield College, so I suggested we change our elementary stats course to a sports stats course, teaching the same material in a sports context. At the time, there were no textbooks with 100% sports examples and problems. So I decided to write the book for the class.

The book that follows is the one I wish I had in college. Because of the focus on sports, I had an added benefit. I was able to cover more statistics than a college-level stats class. I added chi-squared and ANOVA, and we are able to do confidence intervals and hypothesis tests for one and two proportions, and one and two sample means. The other benefit was the ability to teach the class in a computer lab, where we use Minitab for every analysis we do.

Since I was able to design the course around the book, Springfield College now has one of the first college-level stats classes following the guidelines developed by the American Statistical Society in 2003. The Guidelines for Assessment and Instruction in Statistical Education (GAISE) focus on introductory college courses, like ours. The goal is for students to be statistically literate upon completion of their course. The GAISE recommendations are:

- Emphasize statistical literacy and develop statistical thinking;
- Use real data;
- Stress conceptual understanding rather than mere knowledge of procedures;
- Foster active learning in the classroom;
- Use technology for developing conceptual understanding and analyzing data;
- Use assessments to improve and evaluate student learning.

At the conclusion of the course, my goal is to use the above guidelines to insure this class produces statistically educated students through development of statistical literacy and ability to think and interpret statistically. I hope you enjoy your journey through the amazing world of sports statistics!

Carl Fetteroll
Springfield College, Math, Physics, Computer Science Department
Springfield MA

PART 1
Descriptive Statistics

*Brady. Brees. Manning.
Who is the best quarterback?*

PREVIEW: EXPERIENCE SPORTS STATISTICS IN ACTION

The NFL season was at its half-way point in 2009 when its three marquee quarterbacks respectively led their teams to victories. Tom Brady, Drew Brees and Peyton Manning are familiar names even to those who do not follow football, and in week 9 of the 2009 NFL season they did not disappoint. Tom Brady and the New England Patriots beat the Miami Dolphins 27-17. Brady was 25 for 37 and 332 yards, with one interception and one touchdown. Drew Brees helped beat the Carolina Panthers 30-20 by going 24 for 35 and 330 yards, and he also had one pick and one touchdown.

Who had the better day? They both won and they both threw for over 300 yards. The rest of their stats were almost identical. That same night, Peyton Manning led the Indianapolis Colts to a come-from-behind-miracle victory over the Houston Texans to keep his team undefeated. The Colts won 20-17, and Manning was 34 for 50 for 318 yards with, you guessed it, one pick and one touchdown!

Those were three of the top performances of the week, but who had the best game? It's an easy question to answer when athletes go head-to-head in the same competition. For instance, at the 2010 Massachusetts

State Track and Field Championships, three girls broke the state record in the 400M hurdles, so all had phenomenal performances. But they did it in the same race, so it was easy to say who had the best day. With Brady, Brees and Manning, they were in different cities and playing different teams. How do we rank their performances?

The NFL has come up with a statistical method for measuring the performance of quarterbacks. It's called the passer rating, popularly known as the quarterback rating. (In college it's also called passing efficiency.) It's made up of 4 components:

$$C = \left(\frac{COMP}{ATT} - .3 \right) \times 5$$

$$Y = \left(\frac{YDS}{ATT} - 3 \right) \times .25$$

$$T = \left(\frac{TD}{ATT} \right) \times 20$$

$$I = 2.375 - \left(\frac{INT}{ATT} \times 25 \right)$$

where
   ATT = Number of passing attempts
   COMP = Number of completions
   YDS = Passing yards
   TD = Number of touchdown passes
   INT = Number of interceptions

If any result is greater than 2.375, it is set to 2.375. If any result is a negative number, it is set to zero.

The calculations are then plugged into the formula:

$$\text{QB rating} = \left(\frac{C + Y + T + I}{6} \right) \times 100$$

So Tom Brady that day would be calculated as follows:

$$C = ((25/37) - .3) \times 5 = 1.88$$
$$Y = ((332/37) - 3) \times .25 = 1.49$$
$$T = (1/37) \times 20 = 0.54$$
$$I = 2.375 - ((1/37) \times 25) = 1.70$$

$$\text{QB rating} = ((1.88 + 1.49 + 0.54 + 1.70)/6) \times 100 = 93.52$$

Brady's quarterback rating for the game was 93.52. (For comparison, quarterback ratings go from 0 to 158.3 in the NFL.)

Try it! (primecomputing.com has a Quarterback Rating Calculator if you want to play around comparing quarterbacks.) Then calculate the ratings for Manning and Brees and you should get 83.6 for Manning and 96.1 for Brees. So who had the better day? Drew Brees with his 96.1 surpassed Manning and just edged out Brady. Manning's rating was hurt because his completion percentage was lower, and Brees edged out Brady because he accomplished basically the same results but with two less attempts.

So was Brees the best quarterback that Sunday? Actually, no. Jay Cutler, in a loss to the Cardinals, was 29/47 for 369 yards and had 3 touchdowns and 1 interception. His three touchdowns pushed him to a 98.6 rating, the highest for a quarterback in the NFL in week 9.

Be careful how you use statistics though; don't rely on just the statistics. Use statistics to guide you in interpreting the data. Cutler had the highest rating, but his Bears suffered a devastating loss to the Cardinals, a game they should have won. Manning was trying to keep his team undefeated through 9 weeks and had 50 attempts because of the miracle comeback needed to beat Houston. The Patriots were trying to show the Dolphins that they were still the beast of the East. And Brees had his Saints undefeated, poised for a showdown the following week with Indy. Statistics provide us with a tool to help us interpret the game.

The quarterback rating is fine for distinguishing between performances in terms of numbers, but always ask yourself is there something else? Why did Manning throw 50 times? Sports statistics is great for giving us numbers, but don't stop there, we need to analyze and interpret those numbers; don't just accept them at face value. Use them as your starting point in your journey into the world of sports statistics.

QUARTERBACK RATING EXERCISES

Look at this weekend's games in the NFL. Pick two quarterbacks you think will have high quarterback ratings. (Remember, in the NFL quarterback ratings go from 0 to 158.3) Or your two favorite quarterbacks. Or the quarterbacks for your two favorite teams. Pick them now, before the games are played. That will make it more fun. After the games, compute the ratings for each quarterback. Their stats will be at NFL.com. NFL.com will also have their ratings calculated so you can double check your numbers. On Monday turn in your ratings for each quarterback (and show all your work) and you're reasoning as to why one was ranked higher than the other.

Do the same assignment as in Exercise 1, but for two college quarterbacks. For college you'll need a different formula. The NCAA calculates their quarterback rating differently from the NFL. Both use the same stats, completions, touchdowns, interceptions and average yards, but the NCAA system (which most high schools also use) uses a different formula to calculate the rating:

QB rating =

$((8.4 \times YDS) + (330 \times TD) + (100 \times COMP) - (200 \times INT))/ATT$

(The scale for NCAA goes from −731.6 to 1261.6. To put that in a better perspective, the 100 career leaders ratings are from 175.62 to 145.30. The weights (8.4, 330, etc.) were chosen in 1979 so that an average passer would have a rating of 100. Because of rule changes and improved passing, 100 is now a poor passer.)

Comment on the differences in ratings between your two college quarterbacks. Why was one ranked higher than the other?

In 2015, Springfield College (MA) had one of the top rushing offenses in Division III, rushing for 301.3 yards per game, which was seventh in the country. Clearly the Springfield College team does not rely on passing. But a quarterback's rating depends on passing.

Calculate the quarterback rating for Springfield College on a random game of this year. To get the data, Google "springfield college ma football box score". (The Springfield College Pride and Prestosports sites are good choices.) Go to athletics, pick football, click more and pick statistics, then select game log.

As an example, here's the box score for Western New England vs. Springfield College, 9/2/2016.

Springfield

Passing	C-A	YDS	LG	TD	INT
Jake Eglintine	0-1	0	0	0	0
Cory DeSimone	0-1	0	0	0	0
Cam Cooper	0-0	0	0	0	0
Jedi Haynes	1-3	0	0	0	0

Use the formula in Exercise 2.

Does the quarterback rating reflect the type of offense they run? How useful is the quarterback rating in this case?

Turning numbers into data.
Is it magic?

PREVIEW: NUMBERS DO NOT EQUAL DATA. WE NEED CONTEXT

Defining data

30, 40, 93, 95, 142, 155, 175.

Numbers. Seven numbers specifically. Alone, they are just numbers. But do they have any meaning? We need context to understand their meaning. OK then, here is the context:

NCAA DIV 3 X-Country New England Regional Qualifier
USM @ Twin Brooks - Cumberland, ME
Saturday November 14, 2009
11:00 AM (Women's 6000 Meter)
Women's Individual Results

These seven numbers represent the seven women on the Springfield College cross country team who ran in the regionals. The numbers are their scoring place (i.e., Amanda DiPaolo of Springfield College finished in 30th place, and therefore scored 30 points for Springfield College). The top 5 runners for each team score, so Springfield's team score was 400 (30 + 40 + 93 + 95 + 142), which put them in 13th place out of 47 teams competing.

So alone, 30, 40, 93, 95, 142, 155, and 175 have no meaning. But now they make sense. They tell a story. The numbers have become data.

Data is the most important concept in statistics, never mind sports statistics. We will use data every day in class, just as we do in our daily lives. Turn on ESPN and be amazed at the data they dig up and present. Fantasy baseball and football are all about data. Read a box score one morning from a late night west coast game, and it is as if you watched the game. You can visualize the game taking place as you scan the box score. Look at the amount of data needed to calculate the quarterback ratings in Chapter 1. And the ratings themselves are more data for us to analyze and dissect. You can keep the entire lunch-table conversation going with one easy question: Who was the best quarterback of the first decade of the 21st century, Tom Brady or Peyton Manning?

Defining statistics

Now we can define sports statistics. This course is not about memorizing. I will never ask you to define sports statistics, or calculate a quarterback rating without giving you the formula. This course is about finding data and using data to analyze and interpret events in your life. We use sports examples in this class because we have all played sports and we can understand the examples.

You can apply these techniques to the rest of your life; the skills you learn here are transferrable to everyday problems. And the biggest skill is developing the ability (or the confidence) to ask questions. I will show you how to do the statistics, and you will get an answer. But then ask yourself why did that happen? Why was Drew Brees ranked higher than Peyton Manning? Analyze the data. Interpret it. In this class, we may have 25 different interpretations, so don't worry if your interpretation disagrees with the teacher's. (It probably will!) The only answer that would be wrong is the blank one. So, process and don't worry. We will hear everyone's interpretations and learn from them.

Given that, let's define statistics. **Statistics** is a branch of mathematics that collects, organizes, analyzes and interprets data. Sports statistics specifically deals with **COLLECTING, ORGANIZING, INTERPRETING** and **ANALYZING** sports data. We should feel comfortable with the concept of data now, but how about collecting, organizing, analyzing and interpreting the data? Here is another example of numbers:

4, 6, 7, 9, 18, 19, 28

To use these numbers in sports statistics, we need a context. (Remember, numbers + context = data.) So, to give the men their due, here we go:

2011 NEWMAC Cross Country Championships
October 30, 2011
Franklin Park, Boston MA

Event 1 Men 8k Run CC

		Name	Yr	School	Final	Pts
1	234	Dan Harper	SR	MIT	25:28	1
2	242	Roy Wedge	SO	MIT	25:34	2
3	239	Stephen Serene	SR	MIT	25:45	3
4	254	Ryan O'Connell	JR	Springfield	26:07	4
5	236	Ben Mattocks	SR	MIT	26:10	5
6	248	Brian Fuller	SR	Springfield	26:21	6
7	258	Anthony Salvucci	JR	Springfield	26:23	7
8	235	Allen Leung	FR	MIT	26:26	8
9	256	Zach Pietras	JR	Springfield	26:30	9
10	226	Trevor Siperek	SR	Coast Guard	26:33	10
11	204	Gary Ezzo	SO	Coast Guard	26:43	11
12	183	Mark Gulesian	FR	Babson	26:52	12
13	240	Logan Trimble	JR	MIT	26:56	13
14	206	Joe Hill	SR	Coast Guard	26:57	14
15	223	Kevin Obrien	SR	Coast Guard	26:59	15
16	232	Andrew Erickson	SR	MIT	27:00	16
17	238	Eric Safai	SO	MIT	27:03	
18	185	Nathan Kolman	SO	Babson	27:06	17
19	255	Matthew Peabody	SR	Springfield	27:10	18
20	241	Matt Weaver	SR	MIT	27:12	
21	246	David Birdsall	SR	Springfield	27:22	19
22	237	Jay McKenna	SO	MIT	27:25	

		Name	Yr	School	Final	Pts
23	276	Dominic Gonzalez	SO	WPI	27:31	20
24	231	Karl Baranov	FR	MIT	27:34	
25	233	Kris Frey	FR	MIT	27:34	
26	288	Andrew Zayac	FR	WPI	27:37	21
27	284	Lucas Smith-Horn	SR	WPI	27:40	22
28	271	Michael Richard	JR	Wheaton (MA)	27:41	23
29	188	Andrew Oram	SR	Babson	27:44	24
30	217	Myles McCarthy	SR	Coast Guard	27:46	25
31	214	Jordon Lee	SO	Coast Guard	27:49	26
32	283	Shane Ruddy	JR	WPI	27:50	27
33	259	Matthew Scully	JR	Springfield	27:51	28
34	192	Jake Williams	JR	Babson	27:52	29
35	225	Devin Quinn	SR	Coast Guard	27:54	30
36	221	Joseph O'Connell	FR	Coast Guard	27:54	
37	224	Bradley Pienta	SO	Coast Guard	28:02	
38	182	Brian deLeon	SO	Babson	28:05	31
39	274	Scott Burger	JR	WPI	28:10	32
40	190	AJ Skains	SO	Babson	28:11	33
41	222	Ryan O'Neil	SO	Coast Guard	28:17	
42	273	Josh Brodin	JR	WPI	28:18	34
43	219	Brian Musard	FR	Coast Guard	28:19	
44	194	Nathan Buck	SO	Clark (MA)	28:20	35
45	191	Mike Smith	SR	Babson	28:20	36
46	184	William Hallock	SO	Babson	28:24	
47	281	Brendan McKeogh	FR	WPI	28:26	37
48	181	Eduardo Arechabala	SO	Babson	28:36	
49	207	Stephen Horvath	FR	Coast Guard	28:40	
50	251	Corey Hamel	SO	Springfield	28:40	
51	263	Erich Voelker	SO	Springfield	28:41	
52	269	Karl Mader	JR	Wheaton (MA)	28:49	38
53	216	James Martin	FR	Coast Guard	28:56	
54	282	Hunter Putzke	SO	WPI	28:59	
55	250	James Gornell	SR	Springfield	29:00	
56	277	Rob Hollinger	SO	WPI	29:03	
57	198	Joe Kennelly	SR	Clark (MA)	29:04	39
58	189	Chris Pierce	JR	Babson	29:15	

		Name	Yr	School	Final	Pts
59	245	Marck Bashaw	FR	Springfield	29:16	
60	218	Matt Monahann	SO	Coast Guard	29:17	
61	275	Sean Dillon	SR	WPI	29:20	
62	280	Greg McConnell	SR	WPI	29:24	
63	278	Joseph Kelly	FR	WPI	29:26	
64	249	Joe Geurds	FR	Springfield	29:28	
65	287	Keegan Westwater	SO	WPI	29:31	
66	205	Thomas Heikkinen	JR	Coast Guard	29:33	
67	187	Leo Martinez	SR	Babson	29:35	
68	202	Alex Cropley	SR	Coast Guard	29:49	
69	272	Bradford Bailey	JR	WPI	29:53	
70	228	Drew Stafford	SO	Coast Guard	29:58	
71	197	Connor Joyce	SR	Clark (MA)	30:00	40
72	201	Andrew Cook	JR	Coast Guard	30:01	
73	195	Rob Gammel	SO	Clark (MA)	30:03	41
74	252	Tyler Leahy	FR	Springfield	30:05	
75	213	Jon Lash	FR	Coast Guard	30:15	
76	247	Scott Bushey	FR	Springfield	30:19	
77	268	John Green	FR	Wheaton (MA)	30:21	42
78	285	Michael Szkutak	SR	WPI	30:23	
79	286	Mark Vanacore	SR	WPI	30:25	
80	203	Patrick Dreiss	FR	Coast Guard	30:26	
81	220	Bradley Nelson	FR	Coast Guard	30:28	
82	193	Scott Worth	JR	Babson	30:32	
83	212	Zachary Kearney	FR	Coast Guard	30:33	
84	265	Sean Astle	SR	Wheaton (MA)	30:36	43
85	260	Mike Sherry	FR	Springfield	30:38	
86	229	John Tabb	SO	Coast Guard	30:45	
87	266	Harry Bachrach	FR	Wheaton (MA)	30:56	44
88	209	Erick Jackson	FR	Coast Guard	30:57	
89	200	Kyle Sullivan	FR	Clark (MA)	31:06	45
90	208	Ryan Hub	FR	Coast Guard	31:43	
91	186	Michael Liachowitz	SO	Babson	32:28	
92	261	Dan Sugar	FR	Springfield	32:35	
93	227	Zachary Speck	JR	Coast Guard	32:40	
94	199	Sam Morrison	JR	Clark (MA)	33:18	46

		Name	Yr	School	Final	Pts
95	270	Christopher Panzini	SO	Wheaton (MA)	34:00	47
96	264	Nathan Watson	FR	Springfield	35:00	
97	262	Robert Sullivan	FR	Springfield	35:17	
98	267	James Court	SO	Wheaton (MA)	36:00	48
99	253	Christopher Malia	SR	Springfield	37:10	
100	196	Thomas Hurlburt	SR	Clark (MA)	38:20	49

First COLLECTING ...

The numbers above correspond to the scoring places of the Springfield College men's cross country team at the 2011 NEWMAC championships. Here is how the data was **COLLECTED**:

The NEWMACs were held at Franklin Park in Boston and hosted by MIT. All 7 men's teams from the NEWMAC ran their varsity and junior varsity runners together in one race. When each runner crossed the finish line, he was given a Popsicle stick with a number on it. The number corresponded to his place. (The first runner to finish got number 1, the second runner number 2, etc.) The runners then went to the scorer's table and an official **COLLECTED** all the sticks, noting the runner's name and school. When all the runners on all eight teams had turned in their sticks, the data **COLLECTION** phase was complete. The official now had a list of finishers that looked like this:

Then ORGANIZING ...

Now the **ORGANIZATION** phase began. The official's job was to certify who had won the conference championship. He took all the runners and organized them by school. He **ORGANIZED** the data by grouping all the runners with their teams.

Then ANALYZING ...

The **ANALYSIS** portion involved adding up the team scores. That was accomplished by adding together the top five places for each team.

Finally INTERPRETING ...

The officials could **INTERPRET** the scores by looking for the lowest team score. The team with the fewest points is declared the winner. And the team with the second fewest points is the runner-up. The order of finish is listed below:

Team Scores

	Team		Total	1	2	3	4	5	6	7
1	**MIT**		19	1	2	3	5	8	13	16
	Total Time:	2:09:23.00								
	Average:	25:52.60								
2	**Springfield**		44	4	6	7	9	18	19	28
	Total Time:	2:12:31.00								
	Average:	26:30.20								
3	**Coast Guard**		75	10	11	14	15	25	26	30
	Total Time:	2:14:58.00								
	Average:	26:59.60								
4	**Babson**		113	12	17	24	29	31	33	36
	Total Time:	2:17:39.00								
	Average:	27:31.80								
5	**WPI**		122	20	21	22	27	32	34	37
	Total Time:	2:18:48.00								
	Average:	27:45.60								
6	**Wheaton (MA)**		190	23	38	42	43	44	47	48
	Total Time:	2:28:23.00								
	Average:	29:40.60								
7	**Clark (MA)**		200	35	39	40	41	45	46	49
	Total Time:	2:28:33.00								
	Average:	29:42.60								

In terms of times, each individual runner for Springfield College was only 38 seconds behind MIT. Even though MIT dominated the meet, SC only has to make up 3:08 in the fall. That is 0:38 seconds per runner! Now the recruiting starts. And the training. Can Springfield College close the gap even further? Or will some other team step up?

Variable variables

PREVIEW: THE DIFFERENCE BETWEEN CATEGORIES AND NUMBERS

What is a variable?

When we review the data we have seen so far — that is the women's and men's cross country results from the last chapter — we notice that all the finishing places are different. They vary from one runner to the next. That means they are **variable**. The finishing place for each woman is a variable, because it varied from runner to runner. Their names are variables. Their colleges are variables. Their times are also a variable. Now two runners could have had the exact same time (unlikely, but they could have), so in that case is it a variable? Yes, because time can vary from runner to runner.

How about on the men's side? The varsity and junior varsity (JV) ran in the same race. Are varsity and junior varsity different? Yes! Even though they ran in the same race, the runner's level of competition, or classification, is a variable. It varies from runner to runner; some are varsity, and some are JV.

So our job in sports statistics is to investigate characteristics that vary. More importantly we will ask, "What can the variation tell us? Does the variation *mean* something? Is it *significant*?" We will run tests on the data. The tests will tell us, "Yes! Something important is happening!" Or "No, it's just ordinary everyday random variation."

Luckily for us, there are only two kinds of variables in statistics: categorical and quantitative. So, let us begin by understanding both variables.

Categorical variable

A **categorical** variable describes a category, or a group. If what varies are categories, it is a categorical variable. Like position on a basketball team (guard, forward, center), unit on a football team (offense, defense, special teams), country of origin of a hockey team (USA, Canada, Finland). Here are several other examples of categorical variables.

Categorical variable	Levels or group
Level of competition	Varsity
	Junior varsity
	Freshman
	Club
	Intramural
Class	Freshman
	Sophomore
	Junior
	Senior
	Graduate student
Academic level	Academic All-American
	Not Academic All-American status

OK, sounds good. Any variable that can be broken down into groups or categories is a categorical variable. There are only two types of variables, and the first seems pretty easy. How about the second one?

Quantitative variable

Well, the other type of variable is a **quantitative** variable. And the hint here is the root of the word quantitative: quantity. Yes, a quantitative variable deals with quantities, or numbers. If what varies are numbers, it is a quantitative variable. Like the runners times. 26:32.54 is quantitative. It is a number. The height of each runner is quantitative.

The weight of each runner is quantitative. But varsity is a group. It is categorical. Make sense?

Categorical or Quantitative?

Here are some characteristics from football that vary from player to player, so they are variables. Think about each one and decide as you read them whether they are categorical variables or quantitative variables.

- Position
- Height
- Weight
- Team (offense, defense or special teams)
- 40 time
- Division (I, II or III)
- Number of tackles
- Points scored each game

Let's see how you did. Positions include linebacker, running back, kicker, etc., and they are all categories or groups (no numbers!) so position is a categorical variable. Height (6', 6'2", etc.) are all numbers, so they are quantities, which means height is a quantitative variable. Weight (160 lbs, 190, 225, etc.) is represented by a number, so weight is a quantitative variable. Team (offense, defense or special teams) is a categorical variable. 40 time (4.25, 5.1, 5.99, etc.) is a quantitative variable. Division (I, II or III) is a categorical variable. This one is interesting because I, II and III are numbers, but in this case they are being used to describe a category or group. For instance Division III is a group of colleges that share the same commitment to the student-athlete. Number of tackles (2, 35, 112, etc.) is a quantitative variable (the description NUMBER of tackles, tells us right off the bat). And finally points scored per game (40, 41, 55, etc.) is a quantitative variable.

VARIABLES EXERCISE

There are 9 variables. (Each column is a variable.) List them and identify whether they are categorical or quantitative.

Tour de France Winners (1999-2016)

Year	Winner	Country	Time	Avg speed	Stages	Dist (km)	Starters	Finishers
1999	Lance Armstrong*	USA	91.32.16	40.30	20	3687	180	141
2000	Lance Armstrong*	USA	92.33.08	39.56	21	3662	180	128
2001	Lance Armstrong*	USA	86.17.28	40.02	20	3453	189	144
2002	Lance Armstrong*	USA	82.05.12	39.93	20	3278	189	153
2003	Lance Armstrong*	USA	83.41.12	40.94	20	3427	189	147
2004	Lance Armstrong*	USA	83.36.02	40.53	20	3391	188	147
2005	Lance Armstrong*	USA	86.15.02	41.65	21	3608	189	155
2006	Oscar Periero	SPA	89.40.27	40.78	20	3657	176	139
2007	Alberto Cantadar	SPA	91.00.26	38.97	20	3547	189	141
2008	Carlos Sastre	SPA	85.52.22	40.50	21	3559	179	145
2009	Alberto Contador	SPA	85.48.35	40.31	21	3460	180	156
2010	Andy Schleck	LUX	91.59.27	39.59	20	3642	195	167
2011	Cadel Evans	AUS	86.12.22	39.79	21	3430	198	167
2012	Bradley Wiggins	ENG	87.34.47	39.83	20	3488	198	153
2013	Chris Froome	ENG	83:56:20	40.55	21	3404	198	169
2014	Vicenzo Nibali	ITA	89:59:06	40.69	21	3661	198	164
2015	Chris Froome	ENG	84:46:14	39.64	21	3360	198	160
2016	Chris Froome	ENG	89:04:48	39.62	21	3529	198	174

* stripped due to doping

A picture is worth a 1000 words

PREVIEW: A PICTURE REALLY IS WORTH A 1000 WORDS

Now that we are comfortable with which variables are categorical and which are quantitative, we can turn to the question of how we can best understand them. How we can best explain them to others. Well, the best way is with a picture. After all, a picture is worth a thousand words. And with categorical variables, the two best pictures are a **bar chart** and a **pie chart**.

To create a chart we need data. So, let's **COLLECT** some data. (Remember, our definition of statistics was to collect, organize, analyze and interpret data.) Here is the 2012 Springfield College men's soccer team from the Springfield College website.

We want to use this data to answer the question:

Is this a young team? Or is this an experienced team?

2012 Springfield College Men's Soccer Roster

	Name	Class	Pos	Ht	Hometown	Major
0	Billy Schmid	Fr	GK	5-10	Wethersfield, CT	Physical Education
1	Brett Bascom	So	GK	6-2	New London, NH	Recreation Management
2	Tyler Fletcher	Sr	M	5-11	Middletown, CT	Sport Management
3	Tyler Allen	So	M	5-7	Southbridge, MA	Biology
4	Mike Desroches	So	F	5-4	Enfield, CT	Physical Education
5	Joe LaBella	So	M	5-6	Wethersfield, CT	Recreation Management
6	Danny Amato	Jr	M	5-8	Newington, CT	Physical Education
7	Dan O'Grady	So	D	5-8	Somers, CT	Physician Assistant
8	Scott Morneault	Jr	M	5-9	Bristol, CT	Physical Education
9	Zach Dutter	So	M	5-8	Dallas, PA	Recreation Management
10	Alex Reilly	Sr	M	5-10	Ledyard, CT	Physical Education
11	Collin Smith	Jr	D	5-11	Byfield, MA	Elementary Education
12	Joseph McSpiritt	Sr	D	6-0	Mill River, MA	Physical Education
13	John Mankus	Jr	D	6-0	Vernon, CT	Business Administration
14	Michael Fowler	Jr	F	5-9	Maynard, MA	Sport Management
15	Kevin Nowak	So	M	6-0	Plainsboro, NJ	Applied Exercise Science
16	Scott Saucier	Jr	M	5-9	Derry, NH	Sport Management
17	Logan Murphy	So	F	6-2	Troy, NY	Sport Management
18	Drew Sommer	So	D	5-10	West Chester, PA	Physical Education
19	Drew Vanasse	Jr	M	6-1	Concord, MA	Sport Management
20	David Chessen	Sr	D	6-0	Edison, NJ	Physical Therapy
21	Brian Dunn	So	D	5-8	Fiskdale, MA	Physical Education
22	Ryan Malone	Jr	M	6-2	Chicopee, MA	Sport Management
23	Marco Callisto	Fr	M	5-6	Mamaroneck, NY	Undeclared
24	Ben Marcus	Fr	M	5-8	Wappingers Falls	Athletic Training
25	Ryan McAdoo	Fr	M	5-7	Terryville, CT	Physical Education
26	Austin Vico	Fr	F	5-10	Plymouth, MA	Criminal Justice
27	Nick Kobel	Fr	M	6-1	Webster, MA	Sport Management
28	Alejandro Miguel	Fr	D	5-9	Bellmore, NY	Physical Education
74	Chris Walton	Sr	GK	6-5	Cromwell, CT	Applied Exercise Science
77	Mike Breault	Jr	GK	5-10	Seymour, CT	Physical Education

If we look at the percentage of the team that is Freshman, Sophomore, Junior and Senior that will answer the question for us. So our next step is to **ORGANIZE** the data. We will organize it in a frequency table broken down by class (Freshman, Sophomore, Junior and Senior). The table shows how *frequently* each class shows up on the team.

Here is our frequency table:

Springfield College Men's Soccer

	Freshman	Sophomore	Junior	Senior	Total
Frequency	7	10	9	5	31
Relative frequency	22.6%	32.3%	29.0%	16.1%	100%

The frequency row summarizes how many players on the team are Freshman, Sophomore, etc. I got that from the roster. I simply counted how many players were in each class. I then organized it into a table so it would be easier for me to use. The number of players in each class we will call the frequency.

Next we **ANALYZE**. Let's find what percent each class is of the total. So we need to know the total number of players. I counted all the players on the roster. There were 31 men playing soccer for Coach Siebert in 2012. First, let's just look at the freshmen. There were 7 of them, so 7 of the 31 total players were freshmen. Then I divided 7/31 to get 0.226. That is the **proportion** that are freshmen. Move the decimal point two places over to the right and we get the **percent** who are freshmen, 22.6%. It's also called the relative frequency. That just means how the frequency *relates* to the total.

We have now collected and organized the data. We analyzed it by calculating the relative frequencies. We know what percent of the team is in each class. Finally, we get to **INTERPRET** the data. The SC men's soccer team is predominantly middle classmen in 2012. 32.3% are sophomores and 29% are juniors, so 61.3% are middle classmen (as opposed to 38.7% who are freshmen and seniors). That means they are an inexperienced team in terms of senior leadership, but an experienced team in terms of players who are comfortable in Coach Siebert's system.

That is it! We just completed our first exercise in sports statistics. We collected, organized, analyzed, and interpreted data on the men's soccer team. Was this an important question? Yes, it might be! If you are a recruit looking for a D III school where you can get some playing time as a Freshman, the small number of players who are seniors means only five players will graduate. That does not create a lot of openings.

How about if you were the coach? You would need to coach this team differently because the freshmen may get minimal playing time. Also, the Senior leadership gap needs to be filled, so the coach and assistants need to step in. Maybe the team needs different practices to work on the fundamentals of the offense rather than on specific plays which might enhance a weakness on the opposing team. Remember, the freshmen are still learning the coach's system.

The athletic director might have a question. 31 players means about 8 players per class (31 divided by the 4 class is 7.75 or about 8), but there are only 5 seniors. What is it about the team that has the players not returning for their senior year? Or maybe it is not the team. Maybe it is a college-wide phenomenon. Do all teams have relative frequencies similar to men's soccer?

Picturing data in Minitab: Pies and bars

But let's return to our data. Our original goal was to graph our data in two ways, in a bar chart and a pie chart. While the data we're using is pretty simple, pictures can make it easier to compare the numbers.

We will make our first graph on Minitab. Let's graph the men's soccer team. Click on the start icon in the bottom left of your computer screen, then click all programs and pick Minitab, then select Minitab 17. Or type Minitab into the search box and select it. Do not use Minitab express.

In the box under C1 type Frequency. In the box under C2 type Class. Then in boxes 1-4 under Frequency type 7, 10, 9, 5. In the boxes under Class type Freshman, Sophomore, Junior and Senior.

Then select the sixth option, Graph, from the menu bar across the top. Choose Bar Chart, and from the drop down menu pick "values from a table". That is because you just typed in your values in C1, which is the "table" Minitab is going to use. Keep Simple Graph highlighted and click OK. Then click in the box under Graph Variables once to move your curser to the box, and that will bring up Frequency and Class. Scroll over and double click on Frequency. That moves Frequency into the Graph Variable box. Then click on the box to the right of Categorical Variables and scroll over to the left and double click on Class. You will see Class move to the Categorical Variable box (appropriate because class is a categorical variable).

Then select Labels and type in 2012 Springfield College Men's Soccer. Then subtitle 1 type in "Roster by Class". Subtitle 2 is your name. Click OK.

In the footnote you will answer the questions, Is this a young team? As we saw, no, it is not. It is a predominantly filled with sophomores and juniors. That is your interpretation which will always go in the footnote.

For a Pie Chart, click on Pie Chart and then select "chart values from a table". Click in the box labeled Categorical Variables and then double click on C2 on the left. Then click in box labeled Summary Variables and double click on C1. Then click on Labels, and for the Title type in

Springfield College Men's Soccer 2012. Under subtitle 1 type "Roster by Class" and under subtitle 2 type in your name.

For a Bar Chart repeat the above steps except choose Bar Chart. You will have gotten the two graphs below:

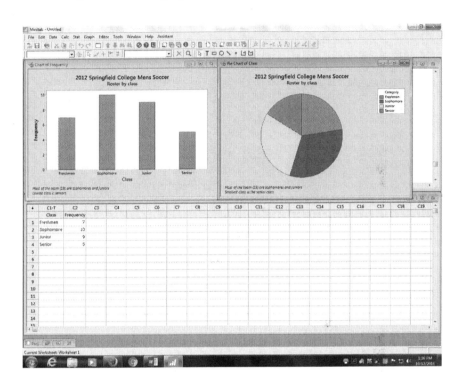

Note: Bar charts and pie charts are interchangeable. Which to use is totally your choice. On a test either one would be right.

To decide which to use, you can ask yourself which makes it easier to see the point you're making with your data. When designing a graph, you have extraordinary power to sway your reader's opinion. People can do most anything with data, twisting it to support their argument. Imagine when an agent goes into a contract negotiation with a general manager. Both will be prepared with binders and binders full of statistics to support their argument for more money (the agent) or less money (the general manager). But there was only one athlete and one set of data, right? Yes, but both sides are doing their utmost to twist the data to support their side of the argument.

Now, let's answer the athletic director's question. Is the low number of seniors in soccer a men's problem or is it a soccer problem? We'll look at the 2009 Springfield College women's soccer team:

2009 Springfield College Women's Soccer Roster

#	Name	Class	Pos	Ht	Hometown	Previous School
0	Hanna MacDougall	So	GK	5-6	Warren, RI	Mt. Hope
00	Erin Greenstein	Fr	GK	5-3	Framingham, MA	Framingham
1	Katie Delude	Jr	GK	5-9	Ludlow, MA	Belchertown
2	Elyse Brightman	So	M	5-4	Franklin, MA	Franklin
3	Amanda Pelkey	Fr	M	5-4	Palmyra, NJ	Palmyra
4	Caitlin Sawczuk	Sr	F	5-4	Plainville, CT	Plainville
5	Amber Thornton	So	M	5-4	Troy, NY	Lansingburgh
6	Allie Davis	Fr	B	5-3	Fairfield, CT	Fairfield Warde
7	Kate Richardson	Jr	B	5-4	Jeffersonville, Vt.	Univ. of Vermont
8	Jen Jahoda	So	M	5-7	Lebanon, CT	Lyman Memorial
9	Sara Dalton	Fr	M	5-5	Schenectady, NY	Mohonasen
10	Ali Dalton	Jr	M	5-3	Schenectady, NY	Mohonasen
11	Jen Courossi	Jr	M	5-3	Methuen, MA	Methuen
12	Gina Calabrese	Jr	B	5-5	East Longmeadow, MA	East Longmeadow
13	Sue Silva	Jr	M	5-5	Ludlow, MA	Ludlow
14	Jackie Moscardelli	So	M	5-2	Weymouth, MA	Weymouth
15	Maria Evans	Sr	F	5-3	Gloucester, MA	Gloucester
16	Kristine Parnell	Fr	M	5-5	Columbia, CT	Lyman Memorial
17	Katie Mantie	So	F	5-3	Wallingford, CT	Lyman Hall
18	Brittany Collins	Jr	F	5-6	Vernon, CT	South Windsor
19	Mary Hogan	Fr	M	5-5	Farmington, CT	Farmington
20	Rory Pogmore	Sr	M	5-7	Lebanon, CT	Lyman Memorial
21	Jess Prencipe	Sr	B	5-8	Andover, MA	Andover
22	Meghan Flanagan	Fr	M	5-5	Averill Park, NY	Averill Park
23	Courtney Price	Jr	B	5-10	Enfield, CT	Univ. of Hartford
25	Laura Hofrichter	Fr	M	5-7	Great River, NY	East Islip

Make a frequency table (remember to include both the frequency and the relative frequency) for the women.

Springfield College Women's Soccer

	Freshman	Sophomore	Junior	Senior	Total
Frequency	8	6	8	4	26
Relative frequency	30.8%	23.1%	30.8%	15.4%	100%

	C1-T	C2
	Class	Frequency
1	Freshman	8
2	Sophomore	6
3	Junior	8
4	Senior	4

Worksheet 1 ***

Interesting, the men have 31 roster spots while the women had 26 in 2009. The men have more sophomores and juniors, while the women have more freshmen and juniors. Both have a big drop-off of seniors. It would be of interest to see if that is still true today.

The reason we looked at the women's data was to see if the low number of seniors was a men's team problem or something more indicative of the athletics program. Well, we see that it is not a men's problem as in this example the women are even worse off in terms of senior leadership with only 15.4% of the team being seniors. Quite a bit less than the 25% estimated!

So if we wanted to look further into the Senior drop off we could check the rosters of other teams. Maybe with kids playing in soccer leagues from the age of four they are burned out by the time they are 21. Perhaps they decide to pursue other activities, like graduating and finding a job.

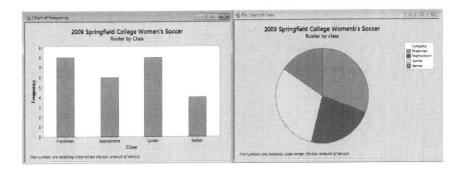

CATEGORICAL VARIABLES EXERCISE

Following is the 2014 Boston Cannons Roster. Create a graph, bar chart or pie chart, of the roster by position.

2014 Boston Cannons Roster

	Name	Position	Height	Weight
28	Brent Adams	Midfield	6' 1" (185 cm)	165 lbs.
85	Mitch Belisle	Defense	5' 10" (178 cm)	195 lbs.
20	Owen Blye	Attack	6' 3" (191 cm)	185 lbs.
12	Martin Bowes	Midfield	6' 2" (188 cm)	200 lbs.
27	Kevin Buchanan	Attack	5' 11" (180 cm)	180 lbs.
22	Craig Bunker	Midfield	5' 9" (175 cm)	185 lbs.
5	Jordan Burke	Goalie	6' 1" (185 cm)	190 lbs.
21	Jim Connolly	Attack	5' 11" (180 cm)	175 lbs.
24	Chris Eck	Midfield	6' 0" (183 cm)	220 lbs.
42	Rob Emery	Midfield	6' 3" (191 cm)	210 lbs.
19	Austin Kaut	Goalie	6' 1" (185 cm)	197 lbs.
1	Will Manny	Attack	5' 9" (175 cm)	160 lbs.
55	Eric Martin	Defense	6' 3" (191 cm)	220 lbs.
72	Scott McWilliams	Defense	6' 3" (191 cm)	205 lbs.
17	Brodie Merrill	Defense	6' 4" (193 cm)	205 lbs.
3	Brendan Porter	Midfield	6' 4" (193 cm)	215 lbs.
99	Paul Rabil	Midfield	6' 3" (191 cm)	220 lbs.
2	Scott Ratliff	Midfield	6' 0" (183 cm)	185 lbs.
43	Jack Reilly	Defense	6' 3" (191 cm)	215 lbs.
11	Matt Smalley	Midfield	5' 10" (178 cm)	180 lbs.
(na)	Erik Smith	Midfield	5' 9" (175 cm)	185 lbs.
41	Mike Stone	Midfield	5' 11" (180 cm)	175 lbs.
77	Kyle Sweeney	Defense	6' 2" (188 cm)	195 lbs.

Time out for testing

PREVIEW: YOUR FIRST STATISTICAL TEST

Now that we've had some experience looking at data, asking questions, finding answers and interpreting those answers, let's take a time out to look at why we're doing this.

Numbers are everywhere. Tom Brady threw 4 interceptions. Patrice Bergeron scored 2 goals. Tuukka Rask made 31 saves.

When the numbers are unusual and unexpected we ask questions. What do they mean? Is this the beginning of the end for Brady? Is Patrice Bergeron the best scorer on the Bruins? Is Tuukka Rask ready to lead the Bruins to a Stanley Cup?

Sports talk radio is kept busy by people who are certain their opinions — their guesses — about what the numbers mean are just as good as anyone else's.

But to make sound decisions — and sound knowledgeable on sports radio — we need more than gut feelings. We want a scientific way to tell us whether the numbers are significant. Does a big jump in interceptions mean something? Or is that in the range of normal random variation?

Statistics has formal processes for answering questions like these. One is called a **hypothesis test**. A hypothesis test is just like it sounds. It tests a hypothesis. A hypothesis is an educated guess. In the past few chapters we asked whether the difference in class sizes on teams was normal variation or whether it meant something was causing seniors to not rejoin the teams. But we couldn't answer that yet. With a hypothesis test we can.

Six steps to a hypothesis test

Our hypothesis tests throughout the course will follow the same six steps.

1. **Null hypothesis**
   Our first step is to ask, "What do we expect the data to look like normally?" What would it look like if there were nothing influencing it? In the soccer roster we looked at, we would expect teams to draw equally from each class, that is 25% freshman, 25% sophomores and so on. We know it won't be this exactly. It will have some natural random variation. But it will be close.

   We call this hypothesis our **null hypothesis** and label it as H_0, h-zero. Think of it as null for zero change. Null equals no. No change. No difference. Nothing going on. So, null hypothesis. That is the hypothesis we test.

   The null hypothesis for the soccer data would be that the team is equally spread between all the classes, that is:

   H_0: The number of Freshmen = Sophomores = Juniors = Seniors.

2. **Alternative hypothesis**
   Our second step is to say something is disturbing our data. Something is influencing it. The differences from the ideal aren't just ordinary random variation. We call this our **alternative hypothesis** and label it H_A. (In some books it's written H_1.)

   The alternative hypothesis for the soccer data would be that the team is *not* equally spread between all the classes, that is:

   H_A: The classes are not equally spread throughout the team.

   The two hypotheses are mutually exclusive. That is, they can't both be true.

   But isn't that redundant? Why have two hypotheses when one is just the opposite of the other? That's a good question! But once we dig into analyzing and interpreting results you will see how helpful it is when we're very clear on what we expect the data to look like ideally and what it looks like when it's not behaving normally.

Here's another head-scratching statistics idea. When we run a test, the result will either reject the idea there's no outside influence — reject the null hypothesis — or fail to reject that there's no outside influence — fail to reject the null hypothesis.

Whoa. What's this fail to reject stuff? Why not just say accept? Because they aren't the same thing! "Accept" means the null hypothesis is true, that the variation is random. But we can't ever know for 100% sure that the variation is normal. When we fail to reject the null hypothesis we're saying don't have enough confidence to say the variation is abnormal. It might be normal. It might not be.

In the soccer team example, we will either reject that variation in classes is random or fail to reject that the variation in classes is random.

It's okay if that's not clear. Just let it sit. It will get clearer the more we work with real data.

3. **Test statistic**
   The third step is the test statistic. Through the magical powers of statistical formulas we can transform an entire set of data into ... a single number! We call it the test statistic. It's a numerical summary of the data. The data can have 5 data points or a million and it can be reduced to just one number. The value of the test statistic will tell us how strong our evidence is our analysis.

   Back in the Dark Ages of my own college days we needed to tediously type all the data into a calculator and crunch the numbers ourselves. Now we can load the whole data set into Minitab, tell it what formula to use and with a press of a button get a test statistic. In this course we will use four test statistics: χ^2 (chi-squared, pronounced kai, like hi).

   We will calculate the different test statistics depending on what we know about the data and what question we're asking. Chi-squared tests are good when we have categorical data. In the next chapter, we'll work with the soccer team data and use the chi-squared test statistic.

4. ***p*-value**
   The fourth step is to calculate the *p*-value. This is the probability that the variation is not random. A lower probability means our variation is less likely to be random.

   So how probable is it that we could expect 7 or 8 of each class on a team of 31 and end up with a team of 7, 10, 9, 5 without something affecting the students' decision to join or leave the team?

5. **Your statistical conclusion**
   What do you conclude from the test statistic and *p*-value? Do you reject the null hypothesis or fail to reject it? There are two possible conclusions:

   • Since the *p*-value is greater than the alpha level, I fail to reject the H_O.
   • Since the *p*-value is less than the alpha level, I reject the H_O.

6. **Your interpretation**
   This is where you become a sports statistician. You will answer what does it mean?

Here are the steps all together. You'll see this list many times in the coming chapters.

>   **Hypothesis Test**
>   1. H_O
>   2. H_A
>   3. Test statistic
>   4. *p*-value
>   5. Your statistical conclusion
>   6. Your interpretation

Don't worry if it's still muddy. It will become much clearer as we work with real data and ask real questions.

Fit fest stats style

PREVIEW: UNDERSTANDING THE GOODNESS OF FIT TEST

Now let's go back to the soccer rosters and look at the data for each class. Since there are four classes, we would expect each class to average 25% of the roster, right? For example, if you had a team with 100 players, you might expect 25 freshmen, 25 sophomores, 25 juniors and 25 seniors. With all else being equal, each class would likely include 25% of the team. We saw in the previous chapter that isn't the case with soccer at Springfield College. We saw that the soccer teams are not equally divided among the four classes.

So is there an issue with soccer at Springfield College or is it just plain randomness? Class is a variable, right? So by definition, it varies. In our above example, if we had 100 players, would a distribution of 24, 26, 25 and 25 be unlikely? No. Sometimes you have better recruiting classes, or more injuries, or academic issues, disciplinary issues, etc. There is almost always a randomness to the distribution. We might be surprised to see 25, 25, 25, and 25 exactly!

So our question for this chapter becomes: Is the class distribution of student-athletes on the Springfield College soccer teams significantly different from the expected frequencies of 25% or are the differences we see due to ordinary random variation?

To answer that question we will use the **chi-squared goodness of fit test**. As was mentioned, this test is good when we have categorical data and want to know how well it fits a theoretical best. We're asking, "Is our data *a good fit* with the theory?" The name of the test actually fits what it does!

Do we have categorical data? We have a count of freshmen, sophomores, etc. Check.

Are we comparing it to real or theoretical data? Our "team" of 25% freshmen, 25% sophomores, 25% juniors and 25% seniors is definitely not real. Again, check.

There are two approaches to a hypothesis test (introduced in Chapter 5), including the chi-squared goodness of fit test.

p-value

The first and easiest way to do a hypothesis test is to find the *p*-value.

There are two new concepts, a *p*-value and an alpha level. Let's tackle the *p*-value first.

The *p* in **p-value** stands for probability. So *p* can be any value from 0 to 1, including 0 and 1. Therefore, a *p*-value of 0.43 is really a 43% probability.

Probability of what? 43% probability of nothing going on, given the data we have. With such a high probability we would fail to reject our initial hypothesis, the null hypothesis, H_0. But how do we decide if the probability is high enough? We compare the probability to an alpha level.

FATROLF
©2017

Throughout this course we will set an **alpha level**, or significance level for the p-value of 0.05. That's our fence. For this course, any p-value we calculate that's above 0.05 means we fail to reject the null hypothesis, e.g., nothing's going on, the differences we see are real differences, but they are due to random variation. Any p-value below 0.05 means we reject the null hypothesis.

Statisticians can tie themselves in knots over choosing a good significance level for the p-value. If you're asking for $1,000 to study whether attitudes on campus discourage women from participating in sports, then a p-value of 0.1 — or even 0.5! — that nothing is stopping them would be good enough reason to do the study. But if the question were whether the bridge would fail or a medicine would kill you, we'd want a p-value that was much smaller, maybe even 0.001!

Since no lives are on the line in sports statistics, being 5% certain nothing out of the ordinary is happening — or 95% certain something is — will be good enough. If you have a job in sports statistics there may be times when you want more certainty or are okay with less certainty. But in this class .05 will work since our purpose is to understand how to work with sports statistics.

Critical value

The second approach to a hypothesis test is to find what's called the **critical value**. If our test statistic (chi squared in this chapter) is less than this critical value then the variation might be ordinary randomness. If the test statistic is above the critical value then we're fairly confident something is going on to affect our data.

In this course we won't use the critical value. But I'll show you how to find the numbers by hand so you understand what Minitab is doing when it hands us a chi-squared test statistic and a p-value.

CHI-SQUARED GOODNESS OF FIT TEST EXAMPLE

Back to the men's soccer team data. Let's finally get an answer to, "Is something going on with men's soccer at Springfield College?"

Remember I mentioned a test statistic reduces a set of data to a single number? In this chapter our test statistic is chi squared. To turn a set of data into a chi squared value we first ask, "How far is each class from the expected value for each class?"

The expected value for each class is the 31 total equally spread between the 4 classes, so 31/4 or 7.75. For the freshmen, how far is 7 from 7.75? It's 0.75. For chi squared it doesn't matter whether the value is smaller or larger. All chi squared cares about is how far the actual value is from the expected value. (In math terms that's absolute value.)

The second step is to square that value. (That also gets rid of the negatives.)

The third step is to divide by the expected value.

Add all those up and that gives the chi-squared test statistic. Here's all the calculations in a table with the chi squared test statistic in the bottom corner. This is relatively easy to do since we only have 4 numbers. It would be an afternoon's work if there were 1000 and we didn't have Minitab.

2012 Springfield College Men's Soccer

Class	Observed freq (obs)	Expected freq (exp)	obs - exp	(obs - exp)^2	(obs - exp)^2/exp
Freshman	7	7.75	-0.75	0.5625	0.07258
Sophomore	10	7.75	2.25	5.0625	0.65323
Junior	9	7.75	1.25	1.5625	0.20161
Senior	5	7.75	-2.75	7.5625	0.97581
Total	31	31	0	14.7500	1.90323

In Minitab we'll enter it the class names and the observed frequencies. It will do all the crunching to return the chi squared test statistic. (You'll see that below.)

Okay we have a chi squared value. What do we do with it? In class we won't do anything with it! It's merely a step that Minitab will do for us. But let's peek behind the curtain at what we used to need to do by hand but now Minitab does for us. We will go find a Chi Squared Distribution Table. (There's one in the Appendix.) Here's a part of it relevant to our question.

Chi Squared Distribution Table (partial)

degrees of freedom	Probability of a larger value of x^2 (p-value)									
	0.995	0.99	0.975	0.95	0.9	0.1	**0.05**	0.025	0.01	0.005
1	---	---	0.001	0.004	0.016	2.706	**3.841**	5.024	6.635	7.879
2	0.010	0.020	0.051	0.103	0.211	4.605	**5.991**	7.378	9.210	10.597
3	0.072	0.115	0.216	0.352	0.584	6.251	**7.815**	9.348	11.345	12.838
4	0.207	0.297	0.484	0.711	1.064	7.779	**9.488**	11.143	13.277	14.860

NON-REJECTION REGION **REJECTION REGION**
Fail to reject the null hypothesis   Reject the null hypothesis

If we cared about different levels of confidence than 0.05 then all the columns would be important. But remember we said for this course we'll choose a significance level of 0.05. So the bolded column is the only one we care about.

But we're still missing bit of information. The **degrees of freedom**. That will always be one less than the number of categories we have. Why degrees of freedom? It comes from asking how many values are free to vary. We know the total (31). Once we know how many students are in any three of the classes we know how many are in the fourth class (31 - the first three classes combined). That means three values are free to vary before they lock in the value of the fourth.

Now go to the partial table above. Find the row that corresponds to our degrees of freedom which is 3. Trace over to the (bolded) 0.05 column. The value there is the critical value, 7.815. If our chi squared value is less than the critical value, it's in the Non-rejection region (colored

red). It's labeled "Fail to reject the null hypothesis" which is what we'd do. If it's greater than the critical value (in the Rejection region), we reject our null hypothesis. Our chi squared value is 1.90323. It is less than 7.815. That puts us in the red area. That tells us we fail to reject the null hypothesis. There isn't enough evidence to suggest the variation is anything other than natural.

That's how we use critical value in hypothesis testing. Which we won't use again. But it wasn't wasted! Return to the table. Trace across the third row to where the chi squared value of 1.90323 would fall. It would be between 0.584 and 6.251. Trace those two upwards into the title bar where the p-values are. The corresponding probability would fall somewhere between 0.9 and 0.1, which is more than the 0.05 value that we set as our fence. The table can't tell us exactly what the value would be. But Minitab can. And it will. So, let's enter the data into the Worksheet.

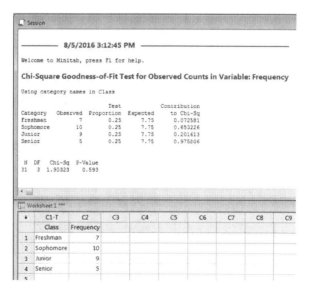

Notice the chi-squared test statistic is 1.90323, which is the same result we calculated by hand! Also the p-value is 0.593. Much easier! But now you have an idea of what Minitab goes off to do behind the curtain.

You just completed the ANALYSIS portion of the collect, organize, **analyze** and interpret the data! Congratulations, right? Before we pat ourselves on the back, we have to interpret the results. Remember, interpretation is the most important step.

Step 5 on the hypothesis test will be: "Since the p-value > 0.05, I fail to reject the H_o." Step 6 is our interpretation. We would say, "There is nothing going on with the men's soccer team. The differences we see in the data are due to random variation." We compared the 7/10/9/8 numbers we found on the roster to the expected numbers 7.75/7.75/7.75/7.75, which are obviously different. But the differences are out to random variation.

Let's use another example to ask the question, "How well does the class make-up of the team fit what we expect, e.g., that each class makes up 25% of the team?" Here's the class make up of the 2010 Springfield College women's basketball team. (F_O is Observed frequency.)

	F_O
Freshman	6
Sophomore	4
Junior	5
Senior	0

Wow! No seniors! We have an unequal distribution of players. Is this due to random variation or is this significantly different? Perform a statistical test to answer this question.

The test needed is a chi-squared goodness of fit test, and it is detailed below. (F_E Expected frequency.)

	F_O	F_E	$F_O - F_E$	$(F_O - F_E)^2$	$(F_O - F_E)^2/F_E$
Freshman	6	3.75	2.25	5.0625	1.350
Sophomore	4	3.75	0.25	0.0625	0.017
Junior	5	3.75	1.25	1.5625	0.417
Senior	0	3.75	-3.75	14.0625	3.750
				$\chi^2 =$	5.533

Adding the last column gave us 5.533, which is our chi-squared test statistic (χ^2). Minitab will do this step for you and also give you a *p*-value of 0.14.

Notes: The expected frequency is 25% (because there are four categories, so each category should have ¼ of the total). So I did 25% times the total number of players, or 0.25 x 15 = 3.75. Since each class is ¼ of the total team, notice the expected frequency is the same for each (3.75).

Also the *p*-value is a probability, so in this case there is a 14% chance that there is nothing going on with women's basketball given those observed frequencies.

Here is how you will answer this question on the exam:

Hypothesis Test
1. H_O: F = S = J = SR
2. H_A: The classes are not equal to each other
3. **Test statistic and its value:** χ^2 = 5.533
4. ***p*-value:** 0.14
5. **Your statistical conclusion:** Since the *p*-value > 0.05, I fail to reject the H_o
6. **Your interpretation:** There is nothing going on with women's basketball. The differences we see are due to random variation.

Investigate the 2016 SC Golf and SC Women's XC teams to determine if there is a good fit between their observed frequencies on their rosters, and their expected frequencies. First, look up the rosters for both teams and enter the data into Minitab as shown below:

	C1-T	C2	C3
	class	observed frequency SC golf	observed frequency womens XC
1	freshman	6	3
2	sophomore	0	6
3	junior	7	4
4	senior	2	6
5			

Then do "Stat", "Table", "Chi-squared goodness of fit" on Minitab, and the results are as follows:

```
Chi-Squared Goodness-of-Fit Test for Observed Counts
in Variable: observed frequency SC golf
                         Test                  Contribution
Category    Observed     Proportion   Expected  to Chi-Sq
freshman        6          0.25         3.75     1.35000
sophomore       0          0.25         3.75     3.75000
junior          7          0.25         3.75     2.81667
senior          2          0.25         3.75     0.81667

N    DF   Chi-Sq    P-Value
15   3    8.73333   0.033
```

Then analyze and interpret the data using the following format:

H_0: F=S=J=Sr
H_A: The classes are not equal
Test statistic and its value: Chi-squared = 8.73
p-value: 0.033
Your statistical conclusion: Since the *p*-value < 0.05, I reject the H_0
Your interpretation: The differences we see are not due to random variation. There is something going on with men's golf at SC.

```
Chi-Squared Goodness-of-Fit Test for Observed Counts
in Variable: observed frequency women's XC
                         Test                  Contribution
Category    Observed     Proportion   Expected  to Chi-Sq
freshman        3          0.25         4.75     0.644737
sophomore       6          0.25         4.75     0.328947
junior          4          0.25         4.75     0.118421
senior          6          0.25         4.75     0.328947

N    DF   Chi-Sq    P-Value
19   3    1.42105   0.701
```

H_0: F=S=J=Sr
H_A: The classes are not equal
Test statistic and its value: Chi-squared = 1.42
p-value: 0.701
Your statistical conclusion: Since the *p*-value > 0.05, I fail to reject the H_0
Your interpretation: There is nothing going on with SC women's XC, the differences we see are due to random variation.

CHI-SQUARED GOODNESS OF FIT TEST EXERCISES

Perform a chi-squared goodness of fit test on Springfield College Women's Softball to determine if the number of athletes, as organized by class, fits the expected model.

First enter the Class, Observed Frequency, Expected Frequency numbers into the Minitab worksheet. Then, after running your chi-squared analysis, answer the following 14 questions on Minitab.

1. What is your null hypothesis (H_o):

2. What is your alternative hypothesis (H_A):

3. Which hypothesis are we testing?

4. Are there differences in the data for softball?

5. What does the null hypothesis attribute those differences to?

6. What test statistic are we using for the Goodness of Fit test?

7. What is the value of your test statistic?

8. What is the alpha (or significance) level?

9. When you perform the test, can you get a p-value > 1?

10. Why or why not?

11. What is your p-value for softball?

12. Statistically, what does your p-value tell you to do?

13. In the context of the problem, what do your results mean?

14. Is there a good fit between your observed frequencies and your expected frequencies?

Perform a chi-squared goodness of fit test to determine if the number of athletes on the Springfield College lacrosse teams, as organized by class, fit your expected model.

Null hypothesis for *both* teams, H_o =

Alternative hypothesis for *both* teams, H_A =

Women's Lacrosse Roster

Class	Observed Frequency	Expected Frequency
Fr		
So		
Jr		
Sr		

Chi-squared =

p-value =

Statistically, what does your calculated p-value tell you to do?

In the context of this problem, what do your results mean?

Men's Lacrosse Roster

Class	Observed Frequency	Expected Frequency
Fr		
So		
Jr		
Sr		

Chi-squared =

p-value =

Statistically, what does your calculated p-value tell you to do?

In the context of this problem, what do your results mean?

Statistics for Every Fan | 45

Does average make sense?

PREVIEW: USING AVERAGE TO DECIDE IF A VARIABLE IS REALLY NUMERICAL

We have already seen there are two types of variables, categorical and quantitative. You have seen categorical variables in practice in the first few chapters. Now we will shift our focus to quantitative variables.

Quantitative variables are numerical. They represent *quantities*. If you can take an average of the data and it makes sense, it is quantitative. For instance, Division I, II, and III are numbers, and therefore numerical, right? Well, does it make sense to take an average of Divisions? No. Try it. II plus III gives us 5, and 5/2 gives us 2.5. We would never average the divisions together.

How about the average of the yards gained for running backs for the Springfield College football team in 2011? Yes! That would make sense. And help us with our interpretation. So yards gained is numerical and therefore, quantitative.

The reason we distinguish between categorical variables and quantitative variables is we will analyze them differently. The easiest way to analyze the data is still by drawing a picture. When we graphed categorical data (in Chapter 4) we could use either a bar chart or pie chart. It didn't make a difference which we used.

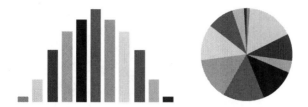

For graphing numerical data we also have two choices of graphs but now it does make a difference which we use. If a **histogram** (a graph like below) has a nice bell shape, that's the graph to use.

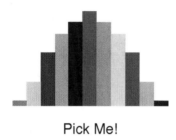

Pick Me!

If the histogram looks like anything else ...

we'll switch to a **boxplot**:

Don't bar graphs and histograms look a lot alike? Yes, they do! Essentially they're the same but are used differently. We use bar charts to compare categorical variables. We use histograms to show how the variables are distributed when they are continuous. The obvious difference — besides gaps between bars in a bar chart and not in a histogram — is that each bar in a bar graph refers to a category.

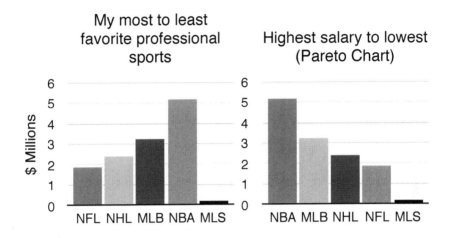

Categories can go in any order you want. Here are the average salaries in professional sports.

In a histogram, the bars refer to numbers. But not just numbers. Groups of numbers. A continuous range. Even though the bars might be labeled with a single number, they represent a range, for example 0-5, 6-10 and so on. The range won't always be integers. It might be batting averages with the range of one bar 0.200 to 0.210 and the next 0.211 to 0.220. It only makes sense if the bars go in numeric order.

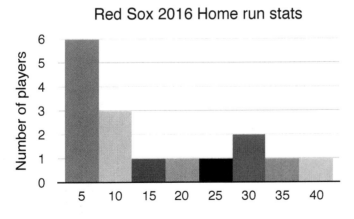

Statistics for Every Fan | 49

A picture is worth a 1000 words (take two)

PREVIEW: EXPLORING MORE GRAPHS TO SUMMARIZE DATA

I mentioned that when quantitative data has a bell shape, we should use a histogram to represent the data. If it's not symmetric, use a boxplot. Real life data will rarely make it that easy for us!

To give you an idea of what symmetrical vs non-symmetrical looks like for real data, I entered the batting averages for all of the qualified hitters in Major League Baseball (MLB) in 2015 into Minitab. Then I had it create three histograms for 2015 American League (AL), National League (NL), and Major League Baseball (MLB) hitters.

Symmetry means we have most of the data in the middle, then the data flows off evenly on both sides. Non-symmetric is everything else (skewed left, skewed right, uniform, bimodal, trimodal, outlier, etc.). Look over the graphs on the next page.

The first histogram is the American League (AL), and you can see it has one peak close to the center and fits *pretty* well under the (red) symmetry curve. In real life we will rarely find data that is perfectly symmetrical, so relative symmetry is enough, and the AL has relative symmetry.

But look at the farthest bar on the right, at 0.340. It's separated from the rest. Go to the top of that bar, then trace over to the left and locate the frequency. That bar has a frequency of 1, meaning one player in the AL is in the group, or bar.

It turns out that bar is the 2015 AL Batting champion, Miguel Cabrera, who hit 0.338 that year. So our histogram has a break between Cabrera and the rest of the batters in AL. That does not automatically make Cabrera an **outlier** (a data point separated from the rest of the data), but in this case, he is an outlier.

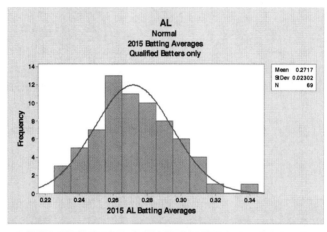

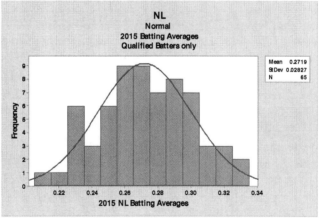

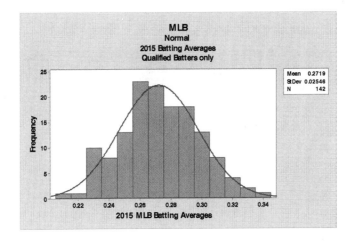

An outlier is found using the 1.5IQR rule. I will show you the calculation when we do boxplots, but for now know that Cabrera is an outlier, therefore we should use a boxplot for the AL.

Now look at the middle graph of the National League (NL). It does not have any outliers, but it does not fit nicely under the curve, mainly because three times it rises, then it falls, making it trimodal. So we will use a boxplot for the NL as well.

But when we combine the two leagues into the MLB graph, it is again *relatively* symmetrical with no outliers, so a histogram will be the good choice here. Notice at 0.230 there is a bump (mode) of 10 hitters, then it goes down again. So the graph is not perfectly symmetrical. But nothing in real life is. If you answer on the exam that you would use the boxplot because of those 10 hitters, I would accept your interpretation as you are literally correct. So you do have some flexibility in interpretation.

Let's try an example. Below is the rushing statistics for the football team. Create a graph of Column 4, Gain (yards gained for the season). If it is symmetric use a histogram. If not, use a box plot. The easiest way to tell is to graph a histogram first, and then you can decide from there.

Springfield College Rushing Statistics for the 2011 Season

RUSHING	GP	Att	Gain	Loss	Net	Avg	TD	Long	Avg/G
Josh Carter	10	196	1549	59	1490	7.6	27	71	149.0
Mike Davis	9	108	641	6	635	5.9	3	37	70.6
Mark Safer	9	61	479	37	442	7.2	5	43	49.1
Austin Bateman	7	54	411	14	397	7.4	2	47	56.7
Brodie Quinn	4	43	252	1	251	5.8	1	93	62.8
Andy Bean	7	24	229	1	228	9.5	2	76	32.6
Joel Altavesta	8	60	224	0	224	3.7	2	16	28.0
Joe Stagliano	1	4	76	0	76	19.0	2	52	76.0
Alex Martin	3	8	66	1	65	8.1	0	24	21.7
Joe Cecala	4	9	48	0	48	5.3	0	14	12.0
Brian Staub	10	10	46	6	40	4.0	1	16	4.0
Franco Bianchi	7	14	50	10	40	2.9	0	16	5.7
Phil Baier	10	2	22	0	22	11.0	0	22	2.2
Rob Merckling	1	1	18	0	18	18.0	0	18	18.0
Ousmane Samb	1	2	13	0	13	6.5	0	11	13.0
Mark Castellano	1	2	8	0	8	4.0	0	6	8.0
Erik Chiarella	1	4	8	0	8	2.0	1	4	8.0
Nathan George	2	3	8	0	8	2.7	0	4	4.0
TEAM	8	6	0	7	-7	-1.2	0	0	-0.9
Total	10	611	4148	142	4006	6.6	46	93	400.6
Opponents	10	417	1996	257	1739	4.2	21	60	173.9

sHAPE, ceNTer & spread

PREVIEW: 3 BASIC QUESTIONS TO UNDERSTAND THE GRAPHS

Before we move onto boxplots, let's look more at how to decide to interpret it. Once you have created your histogram or boxplot, you will want to interpret it. When we describe those graphs, we look at the **shape**, **center** and **spread**.

Shape

Let's talk shape first. As I mentioned, there are two different types of data shapes, symmetric and non-symmetric. This is easiest to see when the data is graphed as a histogram.

Symmetric has a single peak in the middle and is relatively bell shaped.

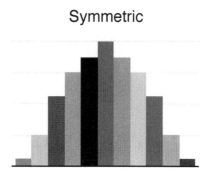

Non-symmetric is ... any other shape! Here are some examples of other, non-symmetric, distributions.

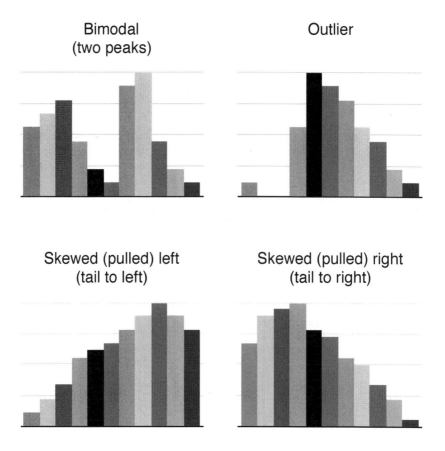

There are even more than these. The data can be trimodal or multimodal. It can be random. Have gaps. Have outliers. Anything that isn't symmetric or fits under a bell-shaped curve.

In terms of the shape, center and spread, that is all we need to know about shape.

Center

What about the center? Well, that is almost as easy. The center is in the center of the graph. It doesn't matter whether the graph is a histogram or a boxplot. The center is the center.

But we use different descriptive stats to describe the center. For symmetric data, represented by a histogram, we use the average of the data. In stats, averaging the data is called the mean.

Non symmetric data pulls the mean, so we cannot use the mean. But a measure of center not affected by skewed data is the median. The median is the middle value when you put the data in order from low to high.

(Minitab will also give us the mode, a third measure of the center. The mode is just the number that occurs most often. I will show you the mode below, but it has limited value in sports examples).

If we have a graph that has a relatively symmetric distribution with one peak in the middle forming a bell-shaped curve we use the mean.

As we defined it earlier, mean is just the statistics word for the average. To calculate the mean by hand, you add up all the numbers and divide by how many numbers you have. But let the Minitab do the drudgery for you. We want to concentrate on interpreting the result without getting bogged down with calculations. You all have access to Minitab. Let it do the grunt work. You can focus on interpreting those numbers.

If we have a graph that looks like anything other than a bell shape we use the median to show the center of the distribution. The **median** is the middle number in a distribution of numbers.

For instance, Tom Brady had quarterback ratings of 59.1, 149.0, and 70.4 in the last three weeks of the 2009 season. His average, or mean, was 92.8, which I got by adding up the three

```
  59.1        divide to    278.5
 149.0        get the       ÷ 3        average =
  70.4        average       92.8        mean
 -----
 278.5

  59.1       put them      59.1
 149.0       in order      70.4       middle number =
  70.4                    149.0          median
```

ratings and dividing by three. His median rating was 70.4, the middle number in terms of putting them in order of lowest to highest, then picking out the middle number.

Both numbers give your center, but remember, we use the mean with a symmetric distribution and the median with everything else. There's one exception. If you have a symmetric distribution *but there is an outlier*, the median is the best way to measure the center of the data. That is because the outlier pulls the mean towards it. The mean is overly excited by the outlier and gravitates towards it. The median, on the other hand, can resist its charms and stays put. So if you have outliers, use the median.

Spread

That takes care of shape and center. Now we tackle the spread of the data. The spread gives us an indication of how wide apart the data is. If the data is bell shaped — symmetrical with no outliers — use the standard deviation to measure the spread. On Minitab, everything will show up on one screen.

Descriptive Statistics: C1

Variable	Mean	StDev	Minimum	Q1	Median	Q3	Maximum	Range	IQR
C1	228.9	377.0	8.0	16.8	58.0	291.8	1549.0	1541.0	275.0

Variable	Mode	N for Mode
C1	8	3

	C1
1	1549
2	614
3	479
4	411
5	252
6	229
7	224
8	76
9	66
10	48
11	46
12	50
13	22
14	18
15	13
16	8
17	8
18	8

If you have data that is uniform, skewed, multimodal, or even symmetrical with outliers, use the interquartile range (IQR) for the spread. What about the basic range you ask? The problem with the range is the max and min might be so large or small that they don't give a good representation of the distribution you are evaluating. So we use the IQR, which is Q3 – Q1.

What are these Q's? Well, it's like a basketball game broken down into quarters. Q1 is the first quartile (the end of the first quarter), Q3 is the third quartile (end of the third quarter). Q3 - Q1 gives us the middle 50% (75% - 25%) of our data.

Another way to look at it is the middle 50% of the data. Most of the time that is much more indicative of the distribution than the range is. The middle 50% of the data is where the bulk of the data lies, and eliminates from our analysis any extremes. So, the IQR = the interquartile range = Q3 – Q1 = the middle 50% of the data. It is that value we use most often to show the spread of the data when it is not symmetrical.

SHAPE, CENTER, SPREAD EXERCISE

Below you will find some data. First I **COLLECTED** my son Danny's 5K times. Then I **ORGANIZED** it into a file in descending order. Now we can **ANALYZE** the data. First, let's look at the data:

Danny's 5K Times

Race	Date	Time
Super 5K	Feb 6, 2005	23:56
Day of Portugal	Jun 12, 2005	24:07
Rhody 5K	Jun 5, 2005	25:32
Mass Senior Games	Jun 25, 2005	25:38
Run for Humanity	Mar 19, 2005	25:44
Norfolk Jingle Bell Run	Dec 11, 2004	25:47
Norfolk Dare	Jun 26, 2005	26:40
Angino Race	Mar 20, 2005	26:46
Gilio Memorial Race	May 28, 2005	27:22
Foxborough Flat 5K	Feb 20, 2005	27:30
Great Bear Run	May 22, 2005	27:30
Medway Lions Road Race	Jun 11, 2005	27:47
Team Hoyt	May 29, 2005	27:58
Bone Density Dash	Jun 4, 2005	28:23
Icicle Series	Feb 19, 2005	28:30

Race	Date	Time
Christopher's Run	May 30, 2005	29:12
Attleboro YMCA	May 7, 2005	29:15
Icicle Series	Feb 5, 2005	29:15
RI Jingle Bell Run	Dec 4, 2004	29:37
Icicle Loop	Mar 26, 2005	29:54
Walter's Run	Dec 19, 2004	29:58
Icicle Series	Feb 26, 2005	29:59
Holyoke Elks	Jun 23, 2005	30:07
Icicle Series	Jan 15, 2005	30:30
Icicle Loop	Apr 30, 2005	30:37
Icicle Series	Feb 12, 2005	30:46
Icicle Loop	Mar 12, 2005	31:26
Icicle Series	Jan 22, 2005	32:51
Icicle Series	Jan 29, 2005	33:44
Icicle Series	Jan 8, 2005	34:50

1. Enter the data into C1. Unfortunately, Minitab does not recognize times, so we have to take 56 seconds and divide 56/60 = .93. Then take the 23 minutes and add the .93 seconds to get 23.93, and enter the 23.93 into minitab.

2. Create a histogram and a boxplot for Danny's 5K times.

3. What is the shape of the distribution of his race times?

4. What is the mean? Median? Standard deviation? IQR (Interquartile Range)?

5. Which measure is appropriate for the center of the distribution? Why?

6. Which measure is appropriate for the spread of the distribution? Why?

7. **INTERPRET** the data.

A box with whiskers?

PREVIEW: A NICE EASY GRAPH BASED ON THE 5-NUMBER SUMMARY

When you have quantitative data, look first at the histogram. It will tell you whether the data is symmetrical or not quickly. If it is symmetrical (or relatively symmetrical), then keep the histogram! If not, for any of the reasons we explored in the last chapter, then switch to a boxplot.

A boxplot is made up of a box that encloses the middle half of the data, with lines (whiskers) extended out left and right to the endpoints of the data (spread). If a data point is an outlier (an extreme value separated from the rest), the boxplot will identify it for you with an asterisk. A boxplot can have multiple outliers (asterisks) or one or none.

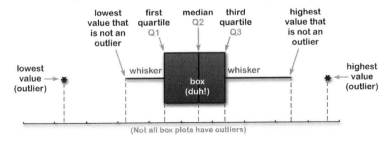

Notice in the graphs on the next page, the AL graph has an asterisk for Miguel Cabrera, highlighting to us that he is an outlier. He is the only one. The second highest average in 2015 was Xander Bogaerts of the Boston Red Sox, who hit 0.320. But his average was close enough to the rest of the AL that he was not an outlier, so the whisker goes out to 0.320 to include Bogaerts, and then stops.

The boxplot has five numbers needed to construct it.

- the lowest value (minimum)
- Q1 (first quartile (quarter))
- median (second quartile)
- Q3 (third quartile), and
- the highest value (maximum)

These five numbers are known as the **five number summary**. The **minimum** is the lowest value in the data set. (In our case, for the AL, Logan Morrison of Seattle, and his whopping 0.225 average.) Q1 is short for the **first quartile**, or the batting average that cuts off the bottom 25% of the data. In 2015, in was Kansas City Royals shortstop Alcides Escobar and his 0.257 average. The **median** is the middle number in the data, or the 50th percentile. Who was right in the middle of the 2015 AL data? Evan Longoria with his 0.270 average for Tampa Bay. Q3 is the **third quartile**, or the batter who hit better than 75% of his fellow players. In 2015 that was Adrian Beltre of the Texas Rangers, and his 0.287 batting average. The **maximum** is the highest average, which brings us back to Miguel Cabrera and his 0.338.

So plot the five number summary (Min, Q1, Median, Q3, and the Max) on a number line, put a box around Q1 to Q3, and then whiskers out to the minimum and maximum. Calculate to see if any of your data points are outliers, and you have a boxplot! Well, what about that calculation for outliers?

Outliers

What makes Miguel Cabrera's average an outlier rather than part of the data?

There isn't a single definition of what an outlier is. It just means his average falls outside a "comfort zone." Statisticians want some way to be alerted to odd-looking data. It's data that may be wrong or so unusual that it will skew their analysis of the rest of the data. In fact there are several ways to identify data that may be odd. We'll just look at the easiest and quickest known as Tukey's range test.

Tukey's test is very simple.

1. We measure the distance from Q3 to Q1. We get that by subtracting the lower value, Q1, from the higher value, Q3. Q3 – Q1 has a special name, it is known as an **Interquartile Range (IQR)** Get it? Inter quartile, between the quartiles, between Q1 and Q3. We will talk more about the IQR later on.
2. Then we multiply the IQR by 1.5. This is called the 1.5IQR rule.

1.5IQR rule = 1.5 × (Q3-Q1)

3. To get the "comfort zone" we add that answer to Q3 to construct a high fence. Then we subtract that same value from Q1 to construct a low fence.
4. Any values above the high fence and below the low fence are outliers.

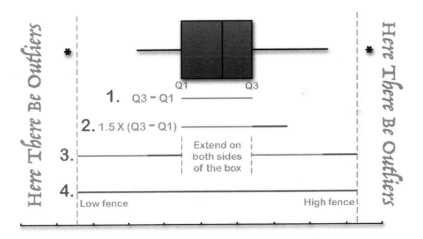

For Miguel Cabrera I calculated the IQR at 0.030 (0.287 - 0.257). Then I multiplied the IQR by 1.5 (0.030) and added that to Q3 to get 0.332. That is my fence. Batting averages below that fence are part of the data. The entire AL was below, except Miggy. His 0.338 was above that fence, making him an outlier. Below is the Minitab output for the boxplot for the AL. Notice Minitab automatically calculates the outlier for us, and labels it with an asterisk. Thanks Minitab!

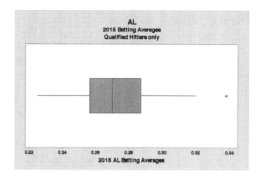

Now back to the Springfield College rushing yards data we looked at in Chapter 8. When you enter Minitab, put your data in C1. (See below.) Then go to graph and pick histogram from the drop down menu. For the option, always keep it simple with fit. Put your curser in the "Graph Variable" box, and click. Then double click on the data value on the left that you want to graph. Click labels and enter a title and subtitle to help your readers to understand the context of your graph. Subtitle 2 will be your name. Then click OK, OK. Here is what it will look like:

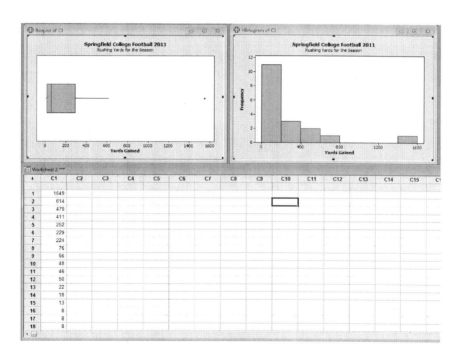

Because the data in the histogram is not symmetrical (it is actually skewed right, remember, it is skewed in the direction of the tail), we need to do a boxplot. Even though I showed them both in the example above, all you need to do is print out the appropriate graph, in this case, the boxplot.

Notice Josh Carter with his 1549 yards is an outlier, a pretty strong outlier. So not only was the data skewed to the right, it had an outlier too. Either one of those facts would have required us to use a boxplot, but both combined confirm the use of a boxplot in this case.

Back-to-back boxplots

PREVIEW: THE GRAPH IS DOUBLY EASY NOW THAT WE HAVE TWO OF THEM

Boxplots give a good picture of quantitative data. Back-to-back boxplots allow us to make a quick and accurate comparison of two sets of quantitative data. For instance, the first decade of the 21st century provided us with two of the greatest quarterbacks of all time, Peyton Manning and Tom Brady. Who was better? Who was a better home run hitter, Babe Ruth or Hank Aaron?

When comparing boxplots we use only the box not the whiskers. Remember, we use a boxplot when the data is not symmetrical. The data outside the middle 50% can be eccentric. Like Robel Kiros Habte's 1:04.95 finishing time in the 100 meter freestyle swim in the 2016 Olympics, nearly 36% slower than the first place finisher. Therefore

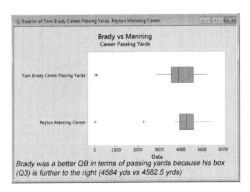

we are leery of the ends, the first 25% and the last 25%. But we are comfortable with the middle 50%. And since the box goes from Q1 to Q3, just like a basketball game, it represents the middle 50% of the data. So when comparing Ruth with Bonds, it will eliminate the high end for anyone who used performance enhancing drugs.

So just compare the boxes, and look to see which box is further to the right.

BACK-TO-BACK BOXPLOT EXAMPLES

For our first in-class example, we will answer the question of which Springfield College soccer team in 2011 was the higher scoring team, the men or the women?

I entered data for the men in C1 and the women in C2. The data I entered (left) is the number of goals scored for each team. Then click on Graph, then enter on boxplot. Then highlight the option of "Multiple Y's Simple". Enter on OK. Click on C1 Men and they will show up in the box for Graph variables. Then click on C2 Women and they will join them. Then click on scale and select the box at the bottom for "transpose values and category scales". Then OK. Then labels and type your Title in the Title box. Then put your name in the Footnote 1 box. Then OK. Then OK again. It will look like this:

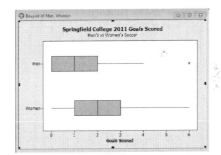

In our example, women's soccer is further to the right because their box is further to the right, so women's soccer is the higher scoring team. On the exam you would answer that in the footnote of the graph on Minitab.

One final note on soccer, notice that both teams had a high of six goals, and each did it once. But because the men were a lower scoring team, the six-goal game shows up as an outlier (an asterisk). The six-goal game for the women is not an outlier, it is part of their data.

Compare the careers of Magic Johnson and Larry Bird in terms of scoring. Look up their total points scored for each season of their careers and create back-to-back boxplots to answer the question, Which player was the better scorer over the course of their careers? Make sure to label your graph, and answer the question in the footnote area of Minitab.

Albert Pujols and David Ortiz are two of the best home run hitters in baseball. Which one though has had the better career? Use back-to-back boxplots to answer the question.

Albert Pujols Career HR	David Ortiz Career HR
37	1
34	9
43	0
46	10
41	18
49	20
32	31
37	41
47	47
42	54
37	35
30	23
17	28
28	32
40	29
31	23
	30
	35
	37
	38

Measuring spread

PREVIEW: THE GRAPH IS DOUBLY EASY WITH TWO OF THEM

Standard deviation. We have seen the term before, but not totally grasped the concept. So let's try: Deviation is a difference. It is how far away from something we are. In this case, we measure how far away from the mean we are. The mean is a measure of the center of the data. Each data point is either greater than (called above in statistics) or less than (below) the mean. Each point *deviates* from the mean. When we look at all the points, we want to find the *average* deviation from the mean. That average is called the standard, so it is the standard deviation from the mean. It is a measure of how far apart the

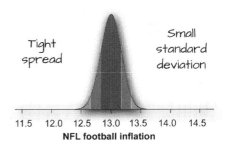

NFL football inflation

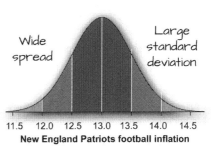

New England Patriots football inflation

data is. Therefore, it measures the spread of the data.

Simple, right? Maybe not! An example will help. In the 2010 pre-season, Tom Brady played in all four games and threw for 67, 85, 273, and 51 yards. Let's find his mean and standard deviation. To find his mean, first we add up all his yards, 67 + 85 + 273 + 51 = 476. Then we divide the total yards by the number of games, so 476/4. That's his average yardage per game. We call that his mean. The mean in this case is 119 yards/game. That is a measure of the center of the data. We want to see how far each of Tom Brady's games deviated from the mean. Some games he was above the mean, some games below the mean. So we create column two below, subtracting each of his yards from his mean of 119.

Yards	Yards - Mean
67	-52
85	-34
273	154
51	-68
476	0

Unfortunately, when we add all the numbers in column two, they add up to 0. How can that be? We can see Brady's yards vary from the mean, some more than the mean, some less. Yet they will always add up to 0 because the mean is like the balance point. It's the point where the numbers to the left balance the numbers to the right. So, to fix that, to get a number we can do something with, we square each number to get rid of the negatives. Those numbers are in the third column.

Tom Brady's 2010 Pre-season Yards

Yards	Yards - Mean	(Yards - Mean)^2
67	-52	2704
85	-34	1156
273	154	23716
51	-68	4624
476	0	32200

Add up the third column and divide by 4, or 32200/4, and get 8050. That is called our **variance**. Now take the square root of the variance, that is the square root of 8050, which is 89.72 yards. We have our standard deviation!

If we had thought about it, we would have expected a large standard deviation because we noticed Brady's yardage varied from 51 to 273, which is a lot. Because of Brady's caliber as a passing quarterback the preseason games may focus on other skills. So we would expect a lot of variation from Brady in the preseason. We would expect his regular season numbers to be more consistent, and therefore have a smaller standard deviation. Let's see.

In the first three games of the 2010 season, he threw for 258, 248, and 252 yards. Remarkably consistent. So we would expect a very small standard deviation as his yardage totals have very little variation. His mean in the regular season in 2010 through three games was 252.7 yards per game. Let's calculate the standard deviation using the same process as above:

Tom Brady's 2010 Yards

Yards	Yards - Mean	(Yards - Mean)²
258	5.33	28.44
248	-4.67	21.78
252	-0.67	0.44
758	0.00	50.67

Adding up the last column (50.67) and dividing by 3 gives us 16.89. Then the square root of 16.89 is 4.11 yards, which is the standard deviation! As we expected, a very small number compared to his 89.72 standard deviation in the preseason.

Empirical Rule: 68/95/99.7

Now we can introduce the 68/95/99.7 rule (or **the empirical rule**). This rule for standard deviation tells us that 68% of all our data values will lie within one standard deviation of the mean. (We just calculated one standard deviation, the 4.11 yards for Brady's regular season passing.) The 95 tells us that 95% of all of our data lies within two standard deviations of the mean. And you can probably guess that 99.7% of the data lies within three standard deviations of the mean.

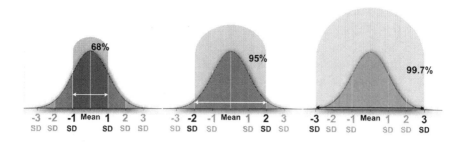

Let's do that with the Brady data. We'll round his numbers to make it easier to see. We'll make the mean 253 and the standard deviation 4. 68% of Brady's yardage values therefore will be between 4 yards below the mean (253-4) and 4 yards above the mean (253+4). So between 249 and 257. 95% of his yardage values will be between 2 standard deviations (4 × 2 = 8) below the mean and 2 standard deviations (again 4 × 2 = 8) above the mean. That is, between 245 and 261 (253 - 8 and 253 + 8). Finally 99.7% of his yardage values will be between 12 yards below and 12 yards above. Which is between 241 and 265 yards.

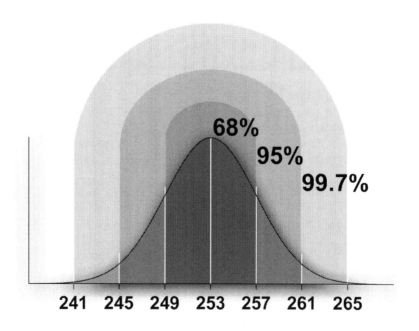

Tom Brady 2010 Passing Yards

This is how we use the concept of standard deviation in statistics.

It will be even clearer with more data. Let's do one more example from MLB. Here is the histogram for MLB in 2016, looking at batting average for all qualified hitters in both leagues. The N = 142 means we have 142 hitters who qualified which is plenty more than the 3 games we had for Tom Brady.

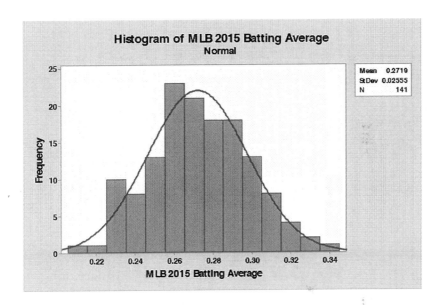

Notice the graph from Minitab also has the mean and the standard deviation! We do not have to calculate it by hand like we did with Tom Brady.

We can also use Minitab to calculate the value without the graph. Put our data in C1, then in the toolbar click on Stat. In the drop down menu, choose "Basic Statistics" and in that drop down menu, choose "Display Descriptive Stats". Double click on C1, which will move C1 to the variables box. Then click on stats. Make sure the mean and the standard deviation are checked. Then hit OK, OK. What you will get is:

Descriptive Statistics: MLB 2016 Batting Average

```
Variable                    Mean      StDev
MLB 2016 Batting Average   0.27194   0.02546
```

Notice the result agrees with our graph! Now we can do the 68/95/99.7 rule for MLB hitters in 2016. The mean goes in the middle and then we add one standard deviation (SD) to the mean, and subtract one standard deviation from the mean.

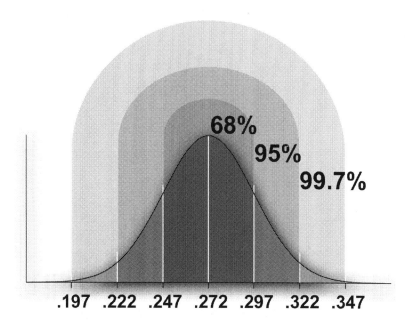

MLB 2016 Batting Averages

That means in 2016, 68% of all MLB hitters hit between 0.246 and 0.297. That is because 68% represents plus or minus one standard deviation from the mean. So I rounded the mean to 0.272, and added the SD of 0.025, which gave me 0.297. Then I subtracted 0.025 from the mean, and that gave me 0.247. So within one SD of the mean just means adding it to the mean once, and subtracting it from the mean once.

Now, when I add the same SD again (0.297 + 0.025), I get 0.322. And when I subtract the SD again (0.247 - 0.025) I get 0.222.

That means 95% of all MLB hitters in 2016 hit between 0.222 and 0.322, or plus and minus two standard deviations. Right? To get up to 0.322, I started at the mean (0.272) and added 0.025 once to get to 0.297, then added it again to get to 0.322. And I subtracted it twice on the other side. And 95% is always plus and minus 2 standard deviations from the mean.

The largest is plus and minus three standard deviations, and that includes 99.7% of the data. Since that is basically just about every major league hitter, we can stop here. So add 0.025 one more time, and subtract it one more time.

So 99.7% of MLB hitters in 2016 hit between 0.197 and 0.347. Hopefully you feel a little more comfortable with the concept of standard deviations now. They just help us to assess how far apart the data is, then we can apply the 68/95/99.7 rule to our data and interpret the result.

STANDARD DEVIATION EXAMPLE

The standard deviation measures the spread of *symmetrical* data, that is how far apart the values are. Specifically, how far the values (x) are from the mean ($\bar{x}$). So the standard deviation measures the standard difference, or the average difference, from the mean.

To calculate the difference is easy. We can do our values minus the mean. And then divide by the total number of values to get the average. But, because some values are negative and some positive, when we add them up we will get zero. So we have to square those values to make them all positive, then we can divide by the total. That gives us a value called the **variance**. The reason it is not the standard deviation is because we squared the differences earlier, so now we have to un-square them (take the square root) to get back to our original units. The formula looks like this:

$$\text{Sample Standard Deviation} = \sqrt{\frac{\sum(x-\bar{x})^2}{n-1}}$$

$$\text{Population Standard Deviation} = \sqrt{\frac{\sum(x-\bar{x})^2}{n}}$$

Let's look at the New England Patriots in 2015. They played 16 regular season games and 2 playoff games for a total of 18 games. I entered the points they scored for each game into C1 into Minitab.

We use the sample formula if we only have a sample of the data, i.e., part of 2016 only, the Patriots first three games only, etc. We use the population if we have everything, the whole season, the entire career, etc.

Then I chose "stat, basic stat, display descriptive stat" to get the mean, which shows on Minitab as 28.33. In the standard deviation formula there is a Greek letter sigma, which means sum. x is our values, in this case the Patriots points scored in 2015. $\bar{x}$ (x bar) is the mean, calculated by Minitab at 28.33. So the formula is telling us for every game the Pats played, take the points scored (C1), subtract the mean (C2), square those differences (C3), and then sum those differences. So when I add up C3, I get 1376.002.

Then I divide by n-1 if I have a sample, and divide by n if I have the population. A sample is when I take part of a season, a population is when I have the whole season. Here I have the entire 2015 season, so I will divide by n. N is the number of games, so I divide 1376.002/18 = 76.4. Finally, remember I have to take the square root of the variance to get the standard deviation, so the square root of 76.4 = 8.74. And that is my answer.

Minitab will do the standard deviation if you select it. Choose "stat, basic stat, display descriptive stats," and then click on statistics and check standard deviation. Now do we graph the Pats with a histogram or a boxplot? To decide, we do the histogram first to see what it looks like. It looks relatively symmetrical, but with two possible outliers. The potential low outlier is the 10 points scored in the final game of the season against the Dolphins. The potential high outlier is the 51 they scored against the Jaguars in week 3. So we do a boxplot to determine if either one is an outlier, so, yes, the Jaguars game in week 3 was. The Dolphins game was not. We, therefore, use the boxplot to graph the Patriots points scored in 2015 because of the outlier.

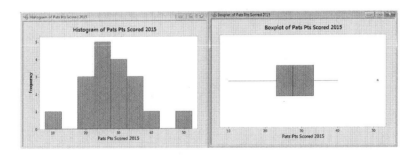

Who had the better day? Z-scores

PREVIEW: A NICE WAY TO FIND WHO HAD THE BETTER DAY

Now we can use standard deviation to help us calculate another statistical method, namely a z-score.

A **z-score** tells us how many standard deviations we are from the mean. A z-score of one means we are one standard deviation above the mean. A score of -1 means we are 1 standard deviation below the mean. A score of 1.6 means we are 1.6 standard deviations above the mean. A score of -0.8 means we are 0.8 standard deviations below the mean. A z-score of zero means we are at the mean. So when we calculate a z-score we are calculating how far something is from the mean. Keep in mind a negative z-score is below the mean, a positive z-score is above the mean.

So we know how to calculate a standard deviation, how about the z-score?

$$Z\text{-}score = \frac{X - \bar{X}}{SD}$$

We take the value we are looking at (like David Ortiz's On-base plus slugging (OPS) in 2016, which is x) and then subtract the mean ($\bar{x}$), then divide by the standard deviation (for all Red Sox batters in 2016, or SD). Is it that easy? Yes!

Z-scores will help us answer the question of who performed better when we have two athletes in two different sports or different positions. For instance, who had a better year in 2015 in the NFL, the passing yards leader for quarterbacks (Drew Brees) or the running back king (Adrian Peterson)? While they both were the statistical leaders for their positions, the positions are completely different. So how can we compare them? With a z-score. The z-score will standardize their results so we can compare. That is, the z-score will tell us how far each is from the average *for their position*. We don't need to guess how far they stand out. A z-score will tell us exactly.

For instance, one standard deviation is better than one half, right? And two standard deviations are better than one. This is because you are further from an average (mean) performance. So if we standardize both values (passing yards and rushing yards) we can compare them apples to apples and see who did better!

By the way, measuring who did better means we have to forget about positives and negatives. We just compare the distance from the mean in either direction. Whoever is farther from the mean had the more surprising result, or the better performance.

For instance, in the results below, Kellie Pennington of Springfield College won the 2012 NCAA Division III 50 Yard Freestyle National Championship in 23 flat. At the same time, Brian Fuller of Springfield College finished third at the NCAA Division III National Championships in Track and Field in the Steeplechase. They were both All-American and had phenomenal performances. Yet who had the better day?

Women's 50 Yard Freestyle

Ln	Pl	Name	College	Time
1	4	Pierce, Laura	The College of NJ	23:31
2	7	Nuess, Morgan	Denison Univ	23:49
3	1	Pennington, Kellie	Springfield College	23:00
4	3	Pavlak, Claire	Emory Univ	23:10
5	2	Raleigh, Christie	Rowan Univ	23:09
6	5	Rosenkranz, Renee	Emory Univ	23:35
7	6	Dobben, Anna	Emory Univ	23:47
8	8	Lukes, Chandra	Univ of Redlands	23:82

Men's 3000M Steeplechase

	Name	Year	School	Finals	Points
1	Nick Kramer	SR	Calvin	8:51.01	10
2	Jack Davies	JR	Middlebury	8:54.61	8
3	Brian Fuller	SR	Springfield	8:56.05	6
4	Michael LeDuc	SO	Conn College	8:59.54	5
5	Anders Crabo	SR	Pomona-Pitzer	9:00.89	4
6	Trevor Siperek	SR	Coast Guard	9:01.53	3
7	Bobby Over	JR	Allegheny	9:03.00	2
8	Jared Brandenburg	SR	Wis.-River Falls	9:04.69	1
9	Stephen Serene	SR	MIT	9:09.23	
10	Andrew Wortham	JR	Bates	9:19.40	
11	Brandon Abasolo	JR	Williams	9:20.78	
12	Marcus Huderle	SO	Carleton	9:22.21	
13	Tyler Morey	SR	Wis.-Oshkosh	9:22.38	
14	Jeremy Kieser	JR	Wis.-Eau Claire	9:34.80	

How can we compare the 3000M steeplechase for men with the 50M freestyle swim for women? They are different distances, different genders, and different sports. Yet we can compare them by standardizing the results so we can look at them apples to apples.

To calculate the z-score on Minitab, go to C1 and enter the 50M Freestyle swim times. In C2 enter the 300M Steeplechase run times. Notice that we enter them in seconds so the program knows that they are quantitative data. Enter your titles in C1 and C2 as in the example below. Then in C3 enter "Swim Z-Score" and in C4 enter "Run Z-Score". Those will be the columns Minitab will use to store the results (the actual z-scores). Then hit calc, and pick standardize. Put your cursor in the box for input column, then double click on C1 and C2. Go to the store results box and double click on C3 and C4. Click OK. You are done! Now we can compare Kellie's performance to Brian's.

	C1	C2	C3	C4
	50m Freestyle	3000m Steeplechase	Swim Z score	Run Z score
1	23.31	531.01	-0.06977	-1.34218
2	23.49	534.61	0.60006	-1.06718
3	23.00	536.05	-1.22337	-0.95717
4	23.10	539.54	-0.85125	-0.69057
5	23.09	540.89	-0.88846	-0.58744
6	23.35	541.53	0.07908	-0.53855
7	23.47	543.00	0.52563	-0.42626
8	23.82	544.69	1.82808	-0.29716
9		549.23		0.04965
10		559.40		0.82655
11		560.78		0.93197
12		562.21		1.04120
13		562.36		1.05419
14		574.80		2.00296

Notice Kellie has a -1.22337. That is her z-score. That score means she was 1.22 standard deviations below the mean for that race. Then look at Brian. He had a -0.95717 z-score. That means his performance was 0.95 standard deviations below the mean, not quite as far as Kellie's was. Therefore, Kellie had the better day! She was farther from the mean. The further you are from the mean, the better you are.

Z-SCORE EXAMPLES

Below are the 2016 Olympic finals for the 3 individual races of Michael Phelps and 2 of Usain Bolt.

Phelps's results are highlighted in blue (C1-C3). Bolt's in orange (C4-C5).

First the average time is found for each race.

Then the average is subtracted from each score. Each score is then divided by the standard deviation. That gives the z-score for each finisher. (C6-C8 and C9-C10.)

Phelps's z-score for the 200 m Individual Medley is -2.13. That means his time is more than 2 standard deviations from the average. That's huge. No one else comes close. (Negative is good in this case since obviously lower times are better.)

But Phelps only had a z-score of -1.01 in the 200M Butterfly. And in the 100M Butterfly he finished second and had -.3. Adding the z-scores and dividing by 3 gives an average z-score of -1.15.

Usain Bolt had z-scores of -1.78 and -1.65 which average to -1.72, much lower than Phelps's average.

So according to the z-scores Bolt had the better Olympics performance.

Sometimes one athlete's z-scores can be negative while the other athlete's z-scores are positive. Do we evaluate them differently? No. Which z-score is further from zero had the better day. But, some students cringe when comparing negatives and positives, so there is an easy trick to fix that. Use absolute values. Since the absolute value is always a positive number you can compare them apples to apples!

For instance, comparing 0.73 and 1.22 is easy, 1.22 is further from zero, so they had the better day.

But how about comparing -0.63 and 0.55? Well, if you can see that -0.63 is further away from 0 than 0.55 is, well you are right! Do not change anything. BUT, if that was not evident to you, take the absolute value first. So the absolute value of -0.63 is 0.63. When you compare 0.63 and 0.55 you can see that 0.63 is further from zero, so they had the better result. But remember when you answer the question on the exam, tell me the -0.63 had the better result because it was further from zero than 0.55.

Absolute values also work if both numbers are negative. Compare -1.44 and -1.49. Take the absolute value and compare 1.44 and 1.49. Much easier! The answer will be -1.49 because -1.49 is further from zero than -1.44.

Here is an example comparing a 50M Freestyle result to a 1M Diving result. Both winners won NEWMAC Championships, but who had the better day?

In Minitab I entered the 50M Freestyle results in C1, then the diving in C2. I named C3 50M Z-scores and C4 1M Diving Z-scores since Minitab will calculate the z-scores for me but it needs a location to store them.

Then I went to "Calc" and clicked on "Standardize." (The z-score is a standardized score that allows me to compare two totally different sports with just one look.) For the input columns I double clicked on C1 and C2 to move over, then for store results in I double clicked on C3 and C4 to move them over. The calculation defaults to "subtract mean and divide by the standard deviation," which is perfect! That is the definition of a z-score. So I do not change that, I just have to hit OK. And the results are shown below:

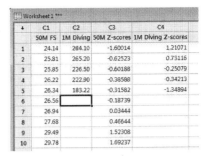

So the winner of the 50 M freestyle race had the better day because their z-score of -1.6 is further from zero than the diver's z-score of 1.21. That is how you will type in your one sentence answer to the question on the exam.

Miguel Cabrera was the AL Batting champ in 2015, with a 0.338 average. The American League average in 2015 was 0.2719 and the standard deviation was 0.02546. Find his z-score and decide if his z-score would be considered unusual.

We do Cabrera's average – AL average, then divide the result by the standard deviation. So 0.338 - 0.2719 = 0.0661. Then 0.0661 divided by 0.02546 = 2.60. So Miguel Cabrera was 2.6 standard deviations above the mean in 2015, which would be unusual because it is more than 2 standard deviations away from the mean.

Compare Ted Williams's batting average in 1941 (his 0.406 that year made him the last major league player to break 0.400!) with Tom Brady's passing yards in 2011 (his 5235 yards in 2011 was a career best) using z-scores and answer the question: Which player had the better season?

Ted Williams
.406 - 0.28646 = 0.11954
Then 0.11954/.03704 = 3.23

Tom Brady Using three games as a starter minimum, I get:
5235 − 2568 = 2667
Then 2667/1468 = 1.82

Ted Williams had the better season because his z-score of 3.23 is farther from 0 compared to Tom Brady's 1.82

To use Minitab, put your data in C1 and C2, then label C3 MLB z-scores and label C4 NFL z-scores. Minitab will calculate the z-scores and put them in those columns. On Minitab go to calc, then standardize (remember, a z-score is a standardized score). Double click on C1 and C2 to put them into the input columns box, then double click on C3 and C4 to tell Minitab where to "store the results in".

That's it! Notice it defaults to subtract mean and divide by the standard deviation, which is what you did by hand above. Then click OK and it will fill in the z-scores for both columns.

Now we can compare MLB with the NFL as we have standardized the values, meaning when we compare the z-scores we are comparing apples to apples. So we can fairly equate Ted Williams and Tom Brady and be comfortable saying that Ted Williams had the better season.

	1946 MLB Batting Averages	2011 NFL Passing Yards	MLB z-scores	NFL z-scores
1	0.406	5038	3.22712	1.66243
2	0.359	5476	1.95832	1.96081
3	0.357	5235	1.90433	1.81663
4	0.340	4933	1.44540	1.61090
5	0.334	4624	1.28343	1.40039
6	0.324	3832	1.01347	0.86084
7	0.322	4177	0.95948	1.09587
8	0.322	3592	0.95948	0.69734
9	0.317	3474	0.82450	0.61695
10	0.314	3610	0.74351	0.70960
11	0.311	4184	0.66252	1.10064
12	0.311	3571	0.66252	0.68303
13	0.307	4051	0.55454	1.01003
14	0.301	3398	0.39256	0.56518
15	0.300	4077	0.36557	1.02775
16	0.299	4643	0.33857	1.41328
17	0.298	2733	0.31158	0.11215
18	0.298	3151	0.31158	0.39691
19	0.297	3091	0.28458	0.35603
20	0.297	3144	0.28458	0.39214
21	0.288	3303	0.04162	0.50046
22	0.284	2214	-0.06636	-0.24142
23	0.283	2164	-0.09336	-0.27549
24	0.282	2497	-0.12036	-0.04863
25	0.280	2753	-0.17435	0.12577
26	0.279	2319	-0.20134	-0.16989
27	0.277	2479	-0.25534	-0.06083
28	0.276	1853	-0.28233	-0.48735
29	0.275	1913	-0.30933	-0.44648

Z-SCORE EXERCISE

Two Springfield College seniors ended their careers here with a bang. Kellie Pennington won another National Championship in the 100M Freestyle at the NCAA National Championships, while Gabby Gaudreault won the Women's Track and Field New England Championship 800M and went to Nationals. Both athletes won their respective races and were crowned champions. So use z-scores to answer the question: who had the better day?

Indianapolis, IN. – March 22, 2014 - Senior Kellie Pennington (Monson, MA) put the exclamation point on her historic career by winning the NCAA Division III National Championship in the 100 freestyle on Saturday night.

	Name	Class	College	Time
1	Pennington, Kellie	SR	Springfield	49.41
2	Bogdanovski, A.	JR	Johns Hopkins	49.66
3	Larson, Nancy	JR	Emory	50.33
4	Yarosh, Hillary	SR	Kenyon	50.94
5	Cline, Jourdan	JR	Kenyon	50.95
6	Pielock, Julia	SR	Connecticut	51.02
7	Kane, Carolyn	FR	Denison	51.23
8	Ternes, Kylie	SR	Johns Hopkins	51.33

Springfield, MA - May 3, 2014 - SC's Gabriella Gaudreault won the 800 m championship on Saturday at the NE Division III Track & Field Championships.

	Name	Class	College	Time
1	Gabriella Gaudreault	SR	Springfield	2:13.18(133.18)
2	Sarah Fusco	JR	Bates	2:14.48(134.48)
3	Cindy Huang	JR	MIT	2:15.44(135.44)
4	Hannah Damron	SO	S. Maine	2:15.66(135.66)
5	Lauren Gormer	JR	Tufts	2:16.58(136.58)
6	Marina Capalbo	JR	Babson	2:17.17(137.17)
7	Chrissy Larrabee	JR	Husson	2:17.24(137.24)
8	Sydney Smith	SO	Tufts	2:17.68(137.68)
9	Sharon Ng	SO	Wellesley	2:18.21(138.21)
10	Nikita Rajgopal	FR	Wesleyan	2:20.05(140.05)
11	Julie Tevenan	SO	WPI	2:20.28(140.28)
12	Ashley Monahan	FR	Westfield St	2:21.32(141.32)
12	Samantha Pomroy	SO	Springfield	2:21.32(141.32)
14	Keelin Moehl	SO	Amherst	2:21.82(141.82)
15	Louise van den Heuvel	SR	MIT	2:21.90(141.90)
16	Addis Fouche-Channer	FR	Middlebury	2:21.92(141.92)
17	Olivia Tarantino	JR	Amherst	2:23.84(143.84)

SPRINGFIELD COLLEGE
Sports Statistics Exam 1 Spring 2015

On the next few pages, you will find copies of previous Exam 1's. Feel free to use them for practice. Good luck!

Drew Brees and Aaron Rodgers are two of the elite QB's in the NFL today. Which one has had the better career based on his passing yards and why? (The stats below are up through the 2015 season. Feel free to expand on that.)

Use back-to-back boxplots to answer the question.

Brees	Rodgers
4,952	4,381
5,162	2,536
5,177	4,295
5,476	4,643
4,620	3,922
4,388	4,434
5,069	4,038
4,423	218
4,418	46
3,576	65
3,159	
2,108	
3,284	
221	

In the 2012 Summer Olympics in London, Arthur Zanetti won the rings competition in gymnastics, and Shelly-Ann Price won the 100M dash in track and field. So they both won gold and were considered the best in the world, but who had the better day? Use z-scores to answer the question and tell me the actual z-scores in your answer.

100M Final

Rank	Country	Athlete	Time
•1	JAM	Shelly-Ann Fraser-Pryce	10.75
•2	USA	Carmelita Jeter	10.78
•3	JAM	Veronica Campbell-Brown	10.81
4	USA	Tianna Madison	10.85
5	USA	Allyson Felix	10.89
6	TRI	Kelly-Ann Baptiste	10.94
7	CIV	Murielle Ahoure	11.00
8	NGR	Blessing Okagbare	11.01

Rings Final

Rank	Country	Athlete	Score
•1	BRA	Arthur Nabarrete Zanetti	15.900
•2	CHN	Chen Yibing	15.800
•3	ITA	Matteo Morandi	15.733
4	RUS	Aleksandr Balandin	15.666
5	RUS	Denis Ablyazin	15.633
6	PUR	Tommy Ramos	15.600
7	BUL	Iordan Iovtchev	15.108
8	ARG	Federico Molinari	14.733

3 The Springfield College Women's Soccer roster is pasted below. Create a graph of the team by class.

Name	Cl.
Brooke Fairman	SO
Holly Ouellette	SR
Lucy Gillett	SR
Lia Gagliardi	JR
Shannon Wells	SO
Nicole Fowler	JR
Abbie McCann	FR
Mikaela Coady	FR
Brooke Hattinger	JR
Julia Cormier	SO
Brianna Messier	SO
Dakota Kelly	FR
Carly Lopergolo	SO
Aubrey Coutu	FR
Mariah Ferrara	JR
Krissy Cicalis	JR
Monique Marcelino	FR
Kimberly Quiles	FR
Jenacee Bradbury	SO
Olivia Bernas	FR
Elizabeth Monsen	SO
Amanda Calzolano	SO
Kelly Haines	SR
Sara Carpentieri	SO
Jessica Miller	JR
Ciara Boucher	FR

Using the same Women's Soccer Roster in problem 3, perform a hypothesis test on the roster by class to measure how well the current roster fits your expected numbers. Identify the 6 components of the test.

 Below is the 2015 Springfield College football team's rushing stats.

Create an appropriate graph of the team's rushing yards (YDS) and interpret the data.

No	Name	Yr	Pos	Rush	Yards
2	Jake Eglintine	SO	QB	156	1089
6	Keith Rodman	SR	FB	118	596
31	Drew Brown	SR	HB	47	343
20	Jordan Wilcox	SO	FB	54	261
19	Tim O'Brien	SR	QB	32	184
26	M. Mastroianni	SR	FB	25	117
38	Cory DeSimone	SO	HB	25	111
24	Andrew Alty	SR	HB	27	104
5	Billy DeLay	JR	HB	6	79
12	Zach Dubiel	FR	QB	15	72
35	Jonny Bianchi	SO	HB	11	64
39	Tyler Hyde	SO	HB	6	24
4	Joe Festo	JR	WR	2	3
1	Kenny Calaj	JR	LB	1	2
21	Nick Valles	JR	HB	1	-5
15	Steve Comee	SO	HB	1	-5
87	Colin Tullson	FR	WR	1	-9
				537	3013
				380	1372

SPRINGFIELD COLLEGE
Sports Statistics Exam 1 Fall 2015

Alex Ovechkin and Sidney Crosby both entered the NHL in 2005, and ever since then have been two of the premier goal scorers in the league. Even last year (2014/15) they ranked 3rd and 5th respectively in the NHL in goals scored, so they continue to play at a high level 10 years into their careers.

Use back-to-back boxplots to answer the question: Who is the better career goal scorer and why?

You can enter the data below or go to hockey-reference.com and copy and paste their career goals scored columns.

Alex Ovechkin

Season	Age	Team	Goals
2005-06	20	WSH	52
2006-07	21	WSH	46
2007-08	22	WSH	65
2008-09	23	WSH	56
2009-10	24	WSH	50
2010-11	25	WSH	32
2011-12	26	WSH	38
2012-13	27	WSH	32
2013-14	28	WSH	51
2014-15	29	WSH	53

Sidney Crosby

Season	Age	Team	Goals
2005-06	18	PIT	39
2006-07	19	PIT	36
2007-08	20	PIT	24
2008-09	21	PIT	33
2009-10	22	PIT	51
2010-11	23	PIT	32
2011-12	24	PIT	8
2012-13	25	PIT	15
2013-14	26	PIT	36
2014-15	27	PIT	28

In the 2016 Rio Olympics, Usain Bolt and his fellow country-person Elaine Thompson both won the 100 meter dash. Which athlete had the better day? Use z-scores to answer the question and tell me the actual z-scores in your answer.

Men's 100-Meter Dash Results

Place	Athlete	Country	Time (seconds)
1	Usain Bolt	Jamaica	9.81
2	Justin Gatlin	United States	9.89
3	Andre De Grasse	Canada	9.91
4	Yohan Blake	Jamaica	9.93
5	Akani Simbine	South Africa	9.94
6	Ben Youssef Meite	Ivory Coast	9.96
7	Jimmy Vicaut	France	10.04
8	Trayvon Bromell	United States	10.06

Women's 100-Meter Dash Results

Place	Athlete	Country	Time (seconds)
1	Elaine Thompson	Jamaica	10.71
2	Tori Bowie	United States	10.83
3	Shelly-Ann Fraser-Pryce	Jamaica	10.86
4	Marie-Josee Ta Lou	Ivory Coast	10.86
5	Dafne Schippers	Netherlands	10.90
6	Michelle-Lee Ahye	Trinidad and Tobago	10.92
7	English Gardner	United States	10.94
8	Christania Williams	Jamaica	11.80

The 2015 Springfield College Women's Gymnastics roster is pasted below. Create a graph of the roster by Class.

2015 Springfield College Women's Gymnastics

Name	Cl	Event	Hometown	Major
Madi Bowen	FR	V, BB, FX	Sunderland, MA	Sports Biology
Briana Kerr	FR	AA	Bolton, CT	Health Science
Linda McAuley	FR	AA	Trumbull, CT	Health Science
Ashley Parchinski	FR	AA	Chester, NY	Physician Assistant
Lauren White	FR	AA	Littleton, MA	Business Management
Gabriela Christ	FR	UB	Buffalo, NY	Sports Biology
Kristin Feliu	FR	V	Elkridge, MD	Physical Therapy
Kelsey Price	FR	BB, FX	Plymouth, MA	Psychology
Jenna Croteau	SO	V, BB, FX	North Kingston, RI	Applied Exercise Science
Marissa DeAngelo	SO	AA	Brookfield, CT	Physical Therapy
Corrine Greis	SO	UB, BB	Metuchen, NJ	Undeclared
Kelsi Levesque	SO	AA	Coventry, CT	Sports Biology
Meri Moreau	SO	V, UB	Attleboro, MA	Sport Management
Olivia Morrell	SO	AA	Southington, CT	Business Management
Nicole Silva	SO	V, BB, FX	Barrington, RI	Sports Biology
Lexie Stiefel	SO	AA	West Chester, PA	Business Management
Nikki Caradonna	SR	UB, BB	Lynnfield, MA	Elementary Education
Abby Clark	SR	AA	Holderness, NH	Rehabilitation & Disability St.
Sarah Libuda	SR	BB	White River Jct, VT	Business Management
Lauren Pocius	SR	AA	Wallingford, CT	Occupational Therapy

 Using the same Women's Gymnastics roster in problem 3, perform a hypothesis test on the roster by class to measure how well the current roster fits your expected numbers. Identify the six components of the test in your answer.

Hypothesis Test
1. H_0:
2. H_A:
3. Test statistic and its value:
4. *p*-value:
5. Your statistical conclusion:
6. Your interpretation:

Google "final 2015 baseball stats" then click on the ESPN link. From the drop down menu next to Season, choose the 2015 regular season (it defaults to postseason). Under ALL MLB choose the "Player batting" link.

Create an appropriate graph for batting average and interpret the data.

SPRINGFIELD COLLEGE
Sports Statistics Exam 1 Spring 2016

Tom Brady and Cam Newton were two of the best quarterbacks in the NFL in 2015. Tom had more touchdowns (36 to 35), fewer interceptions (7 to 10) and a better QB Rating (102.2 to 99.4), but somehow Newton won the MVP. So maybe Newton was better on passing yards?

Use back-to-back boxplots to answer the question: Who was the better quarterback in 2015 based on passing yards, Tom Brady or Cam Newton?

Newton Yards/Game	Brady Yards/Game
175	288
195	466
315	358
124	275
269	312
197	355
248	356
297	299
217	334
246	277
183	280
331	312
265	226
340	267
142	231
293	134

On January 16, 2016, the Springfield College Women's Swimming and Diving team swam against NYU and were led by senior diver Melanie Avdoulos, who won two events. On January 30, 2016, the Men's Indoor Track and Field team dominated the competition, winning Coach Klatka's final home indoor meet of his career. The men were led by Junior Alexander Niemiec, who won the 60 meter dash.

So Melanie and Alex both had excellent meets, but which athlete had the better day? Use z-scores to answer the question and tell me the actual z-scores in your answer.

Women 1 Meter Diving

	Name	School	Time
1	Avdoulos, Melanie	Springfield-NE	265.20
2	Campitelli, Austin	NYU-MR	263.32
3	Pankonin, Ashlie	H NYU-MR	259.51
4	Skaza, Sierra	Springfield-NE	254.56
5	Polutchko, Emily	Springfield-NE	204.16
6	Liu, Claire	NYU-MR	177.38

Men 60 Meter Dash

	Name	Yr	School	Time
1	Alexander Niemiec	JR	Springfield	7.00
2	Samir Williams	FR	UMass Dartmo	7.13
3	Alex Tomcho	FR	Trinity	7.18
4	Michael Hunter	FR	Springfield	7.30
5	Josh Young	JR	Amherst	7.42
6	Tyler Mach	FR	Springfield	7.45
7	Stadtler Thompson	FR	Amherst	7.45

Below is the 2016 Springfield College Women's Lacrosse team. Create a graph of the roster by Class (Cl).

2016 Springfield College Women's Lacrosse Team

	Name	Cl	Pos	Ht	Hometown	Major
0	Gabriella Anderson	Jr	G	5-7	Mahopac, NY	Elementary Education
1	MK Jaeger	Sr	D	5-3	Plymouth, MA	Physical Education
2	Paige Campbell	Jr	A	5-2	Lakeville, MA	Applied Exercise Sci
4	Kelsey Orpin	Fr	M	5-2	Pittsfield, MA	Physical Therapy
5	Taylor Kane	Fr	M	5-6	Winchester, MA	Applied Exercise Sci
6	Emily Mancini	So	D	5-3	Cheshire, CT	Physician Assistant
7	Kristina Krull	Sr	M	5-3	Leicester, MA	Applied Exercise Sci
8	Shelby Corsano	So	A	5-7	Bourne, MA	Physical Therapy
9	Katie Stallone	Fr	A	5-4	Seaford, NY	Occupational Therapy
10	Taylor Black	Fr	A	5-1	North Merrick, NY	Athletic Training
11	Jennifer Ryan	Sr	D	5-8	Webster, NY	Physical Therapy
12	Jacey Miller	Fr	D	4-11	Waterford, CT	Sports Biology
13	Andie Stone	Fr	D	5-8	Burlington, CT	Occupational Therapy
14	Kristen Steidler	So	M	5-4	Somers, CT	Physician Assistant
16	Amanda Picozzi	Fr	D	5-3	Wyoming, RI	Sports Biology
17	Amanda Nusbaum	Fr	M	5-4	Conway, NH	Sports Biology
18	Kayla Schroeher	So	M	5-4	Riverhead, NY	Physical Education
19	Heather Raniolo	Sr	M	5-2	Yorktown Heights, NY	Physical Education
20	Stephanie Hyslip	So	A	5-4	Hanson, MA	Criminal Justice
21	Bianca Raniolo	Sr	A	5-2	Yorktown Heights, NY	Psychology
22	Julia Bireley	So	D	5-2	Franklin, MA	Sports Biology
23	Lilly Barraclough	Fr	M	5-4	Dover, NH	Health Science
24	Helen Dinnan	Fr	A	5-9	Southington, CT	Undeclared
25	Sam Scanu	Jr	D	5-4	Averill Park, NY	Physical Therapy
27	Bridget Thibodeau	Sr	D	5-6	Andover, MA	Elementary Education
28	Ann Mahoney	So	M/D	5-7	Saratoga Springs, NY	Occupational Therapy
32	Allie Goddard	Fr	A	5-9	Acton, MA /	English

Using the same Women's Lacrosse roster in problem 4, perform a hypothesis test on the roster by class to measure how well the current roster fits your expected numbers. Identify the six components of the test in your answer.

Hypothesis Test
1. H_o:
2. H_A:
3. Test statistic and its value:
4. *p*-value:
5. Your statistical conclusion:
6. Your interpretation:

Google "Oakland Athletics 2015 Pitching Stats." Click on ESPN. Copy the final column, ERA, into Minitab.

Then create an appropriate graph for ERA (earned run average) and interpret the data.

SPRINGFIELD COLLEGE
Sports Statistics 🏈 Exam 1 Fall 2016

Last week we witnessed the last regular season baseball games for two icons, David Ortiz and Mark Teixeira. Ortiz is the most productive designated hitter (DH) of all time, ranking number 1 in home runs, RBI's and hits. Teixeira is one of the most productive switch hitters of all time, setting the MLB record for most games (14) with a home run from both sides of the plate.

Use back-to-back boxplots to answer the question: Who was the better career home run hitter?

To find the data, Google "David Ortiz Stats" and choose the baseball-reference link. Scroll down to the HR column.

Then do the same for Teixeira.

In the 2016 Rio Olympics, Nafissatou Thiam of Belgium won the gold medal in the women's heptathlon and was crowned the world's greatest women athlete.

Meanwhile, Usain Bolt completed the triple triple, including another gold medal in the 100M, marking him as once again, the world's fastest human.

So they both won gold medals, but which performance was better? Use z-scores to answer the question: which athlete had the better meet, Thiam or Bolt? Make sure to include their z-scores in your answer.

Google "Olympic 2016 results" and click on the link for ESPN. In the sports box click on "athletics" and scroll down to women's heptathlon, and click on complete results. Then copy the results into Minitab. Then go back and scroll down for men's 100M final, and copy the results into Minitab.

Go to Springfield College's website, and click on athletics. Then click on men's cross country. Create a graph of the team's roster by class.

In question 3, you found some variation in the numbers for men's cross country by class. Is there a statistical difference in the numbers you found?

Perform a hypothesis test to answer the question. Include the 6 components of the test in your answer.

Hypothesis Test
1. H_0:
2. H_A:
3. Test statistic and its value:
4. *p*-value:
5. Your statistical conclusion:
6. Your interpretation:

In 2015, Patrice Bergeron led the Bruins in scoring with 67 points. Create an appropriate graph for Bergeron's points scored and interpret the data.

To find the data, Google "Bruins 2015 stats". Click on the "NHL.com" link. Scroll down and copy the point's column (labeled P). Do not include players with zero points.

PART 2
Predictive Statistics

PREDICTION PERFECTED

Simpson's paradox

PREVIEW: THIS HAPPENS A LOT EVEN THOUGH IT SHOULDN'T!

This unique topic was first discovered by Karl Pearson in 1899, but it was not published until 1951 by mathematician Edward Simpson, so Simpson gets the credit and his name will forever be attached to the concept. A paradox is something that happens, but it should not happen. A real head scratcher. For instance, if I do a triathlon with 2016 Olympic Gold medalist Gwen Jorgenson, and I beat her in the swim, I am faster than her on the bike, I beat her in both transitions, and I beat her on the run, I should beat her in the race! So if somehow she ends up beating me that would be a paradox. Here is another example:

The table summarizes the shooting percentages in the 2001-2002 season in the NBA by Brent Barry and Shaquille "Shaq" O'Neal.

NBA Shooting percentages

Shot type	Brent Barry		Shaquille O'Neal	
2 point shot	237/403	58.8%	712/1228	58.0%
3 point shot	164/387	42.4%	0/1	0.0%
overall	401/790	50.8%	712/1229	57.9%

Notice in the regular season Barry beat Shaq in 2 point shots, and also beat Shaq in 3 point shots. There are no other shots that are part of their shooting percentage (free throws have their own percentage). So this is everything, and Barry beat Shaq in both, so Barry should beat Shaq when we combine them, right? Wrong! Barry finished the season lower than Shaq on shooting percentage even though he was better in all categories of shooting percentage.

That is the paradox. Something happened that should not happen. If you are better in the individual categories, you should be better when they are combined! But Barry was not. That is Simpson's Paradox. Now, why did it happen?

Barry shot almost half of his shots from 3 point range, which are harder to make and therefore have a lower percentage. Shaq took only one shot from beyond the arc (and missed it), so his overall average was hardly affected at all! His average stayed pretty close to his 2 point percentage of 58%, while Barry's got pulled down to below 51% because of the harder 3 point shot. So it was an unequal distribution of numbers, Barry having taken so many more threes than Shaq.

Another example comes from Springfield College student Joseph Brown and from MLB.

Philadelphia Phillies

Player	2006		2007		Combined	
Ryan Howard	182/581	0.313	142/529	0.268	324/1110	0.292
Gregg Dobbs	10/27	0.370	88/324	0.272	98/351	0.279

So Dobbs had a higher batting average in 2006 and 2007, but combined Howard was better. It was because of an unequal distribution of numbers. Howard had the majority of his at bats in 2006 when he hit 0.313, and Dobbs had the vast majority of his in 2007 when he hit 0.272.

And also from Joe:

New York Mets

Player	2008		2009		Combined	
Angel Pagan	25/91	0.275	105/343	0.306	130/434	0.300
Carlos Beltran	172/606	0.284	100/308	0.325	272/914	0.298

So Beltran was better both years, but Pagan was better overall. It was because of an unequal distribution of numbers. Beltran had the vast majority of his at bats in 2008 when he hit 0.284, bringing his average down. Pagan had the vast majority of his at bats in 2009 when he hit 0.306, bringing his average up and ultimately higher than Beltran.

One more baseball example with Dustan Mohr (Twins and Giants) and Darin Erstad (Angels) during the 2003 baseball season. Here we compare their batting averages with runners in scoring position and no runners in scoring position. Take a look:

2003 Baseball Season

Player	Runners		No runners		Overall	
Dustan Mohr	19/97	0.196	68/251	0.271	87/348	0.250
Darin Erstad	9/50	0.180	56/208	0.269	65/258	0.252

So Mohr hit better with runners in scoring position, and with no runners in scoring position. But, Erstad hit better overall. You can see the problem with the overall distribution of the numbers. Erstad only had 50 at bats when he coughed up a 0.180 average. Since it was only 50, it did not pull his overall average down as much as Mohr's did (as Mohr had almost twice as many at bats with runners in scoring position).

And back to the NBA for our last example. These are team (rather than individual) shooting stats from Game 5 of the 2011 NBA playoffs between the Spurs and the Grizzlies:

2011 NBA Playoffs Game 5

Team	2-point shooting		3-point shooting		Combined	
Grizzlies	38/77	49.4%	3/10	30.0%	41/87	47.1%
Spurs	32/63	50.8%	7/22	31.8%	39/85	45.9%

So the Spurs had better 2-point accuracy, better 3-point accuracy, but overall were outshot by Memphis! That is the paradox, the Spurs should have been better overall. But what have we learned from this section? Statistics is a tool in our toolbox, but not the only tool. Sometimes we cannot take the data at face value, we have to dig deeper. We have to find another tool in our toolbox.

SIMPSON'S PARADOX EXERCISE

In 1995,

Derek Jeter was 12 for 45. What was his batting average? _____

David Justice was 104 for 411. What was his batting average? _____

Who was the better hitter and why?

In 1996,

Jeter was 183 for 582. What was his batting average? _____

Justice was 45 for 140. What was his batting average? _____

Who was the better hitter and why?

Combining 1995 and 1996,

Jeter was 195 for 630. What was his batting average? _____

Justice was 149 for 551. What was his batting average? _____

Who was the better hitter and why?

One was better in 1995 and 1996 but the other was better when the years were combined. Why?

In 1997,

Jeter was 190 for 654. What was his batting average? _____

Justice was 163 for 495. What was his batting average? _____

Who was the better hitter and why?

Combining 1995, 1996 and 1997,

Jeter was 385 for 1284. What was his batting average? _____

Justice was 312 for 1046. What was his batting average? _____

Who was the better hitter and why?

One was better in 1995, 1996 and 1997 but the other was better when the years were combined. Why?

Title IX: Thou shalt not discriminate

PREVIEW: FEDERAL LAW MANDATES EQUAL OPPORTUNITIES FOR ALL GENDERS

Title IX of the Education Amendments of 1972 is a United States law enacted on June 23, 1972. The law states that no person in the United States shall, on the basis of sex, be excluded from participation in, be denied the benefits of, or be subjected to discrimination under any education program or activity receiving Federal financial assistance — United States Code Section 20.

Although Title IX is best known for its impact on high school and collegiate athletics, the original statute made no explicit mention of sports.

Three-prong test of compliance

In 1979, the US Department of Health, Education, and Welfare under President Jimmy Carter's administration issued a policy interpretation for Title IX, including what has become known as the "three-prong test" of an institution's compliance.

PRONG ONE - Providing athletic participation opportunities that are substantially proportionate to the student enrollment, OR

PRONG TWO - Demonstrate a continual expansion of athletic opportunities for the underrepresented sex, OR

PRONG THREE - Full and effective accommodation of the interest and ability of underrepresented sex.

A recipient of federal funds can demonstrate compliance with Title IX by meeting any one of the three prongs.

Springfield College has an undergraduate enrollment that is 58% male and 48% female. So, 52% of the athletic opportunities should be for men. Prong I of Title IX says that athletic opportunities should be proportionate amongst men and women. So women at SC should have 46% of the athletic opportunities. Do they? Does Springfield College meet prong one?

Prong I is the easiest to measure, and since a college only has to satisfy one prong of Title IX, if you can satisfy Prong I you are good to go!

To answer the question, go to the Springfield College athletics website, and add up the roster spots for the women's teams and the men's teams separately. That becomes our observed frequency. Right? We just observed the counts, or frequencies, of women and men participating in intercollegiate athletics here at Springfield College.

For the expected frequencies, we look at the language of prong one. These athletic opportunities that you just counted have to be "substantially proportionate" to the Springfield College student enrollment. So for the men, since 52% of the student body is male, we would expect 52% of the athletic opportunities to be open to men. We calculate 52% times the total number of roster spots for men and women combined. We enter that answer for the expected frequency for the men. For the women we enter 0.46 x the total number of roster spots for men and women combined.

To calculate if those numbers are substantially proportionate, we perform a hypothesis test. If the p-value is greater than 0.05, we fail to reject the null hypothesis. The null hypothesis in this case is that there is no difference between the observed and the expected frequencies, or if they are different, it is due to random variation. In the context of this problem, since the p-value is greater than 0.05, we would say that the differences between the observed values and those we expected to see are due to random variation. We would not have enough evidence to say the Springfield College is out of whack and therefore the College's athletic opportunities are substantially proportionate to the undergraduate student body.

If our *p*-value is less than 0.05, we would reject the null hypothesis. Our conclusion would be that we have enough evidence to say the Springfield College is violating prong one of Title IX. That the College is not providing enough athletic opportunities for the underrepresented sex.

If that was the case, it does not mean that Springfield College is in violation of Title IX. Any college or university can satisfy Title IX through compliance with just one of the three prongs, they do not have a legal requirement to satisfy all three.

Let's examine the data in Minitab.

In C1, label it "Observed Counts SC Athletes", C2 will be "Gender" and C3 will be "Undergraduate Proportions"

The observed counts are the roster spots that we counted up for men and women. Gender is the categorical variable, and we can enter Male and Female. The undergraduate proportion is tricky. First Google "percent undergrad female Springfield College". The first link will tell you that we are 48% female (and therefore 52% male). BUT, a percent is not a number. To convert percent to a number, move the decimal point two places to the left. So 48% becomes 0.48. And 0.48 is what we enter in Minitab (and 0.52 for the men).

Then click on Stat, Tables, Chi-squared Goodness-of-Fit Test (One variable). Then enter the observed counts and the categorical variable (C2). But the test defaults to equal proportions, which was fine when we did the first exam. But now we have unequal proportions. (In SC's case, 52% vs 48%, notice they are unequal.) So click on specific proportions and put C3 in that box.

Note: Be careful. On the first exam we did the exact same chi-squared goodness of fit test, but we used the default, equal proportions. When we do chi-squared for exam 2, it is hard to remember to switch to specific proportions because it defaults to equal AND we used the default the first time through. So go slow when you do the test on exam 2, the only change is to use specific proportions instead.

We will do Springfield College in class so we have the latest data. But Boston University is on the next page, and Providence College after that.

TITLE IX EXAMPLE

Boston University was a national power in football at the I-AA level as it was called then, even playing in the national championship game. Then BU dropped football as part of their Title IX efforts to support women's teams. So does BU now satisfy Prong I of Title IX?

I went to the BU website and found all the teams for varsity athletics for women and men, and I added up all the roster spots.

BU Sports 2014

Men	Sport	Women
14	Hoops	10
19	XC	22
0	FH	18
0	Golf	10
26	Hockey	25
50	LAX	30
44	Rowing	75
35	Soccer	27
0	SB	21
26	Swim/Dive	29
10	Tennis	9
46	T&F	56
270	Total	332

So the athletic department looks good so far. They have 62 more roster spots for women. But it turns out the undergraduate enrollment at BU is 60% women and only 40% men. So they should have more opportunities for women. Women should have about 60% of the opportunities. If the women do have about 60%, then BU satisfies Prong I of Title IX. To test that they do, we perform a Chi-squared Goodness of fit test, which we have seen before.

1. H_0: By gender, the proportion of undergrads = proportion of athletes
2. H_A: The proportions are not equal

Now we enter our data into Minitab. In C1 we will put the observed counts. In C2 the gender. In C3 the proportions. With the proportions, Minitab will not accept the percent's, we have to enter the percent as decimals. So just move your decimal point two places to the left to make it into a decimal number.

Important note: Remember it defaults to equal proportions. Click on specific proportions to change this.

Click on "Stat" then "Tables" then "Chi-squared Goodness of Fit" For observed counts, double click on C1. For Category Names double click on C2. For specific proportions, double click on C3. Then OK. The result of the test is here:

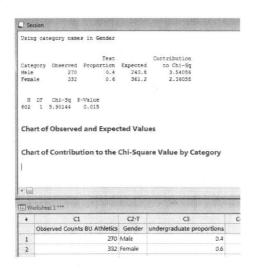

3. **Test statistic and its value:** χ^2 = 5.90
4. **p-value:** 0.015
5. **Your statistical conclusion:** Since the p-value < 0.05, I reject H_o
6. **Your interpretation:** The differences we see in BU's numbers are not due to random variation. BU does not satisfy Prong I of Title IX.

TITLE IX EXERCISES

Providence College has no football program, which means the College has only 204 varsity roster spots for men. The women have 179 roster spots, but make up 56.5% of the undergraduate population, while the men make up the balance (43.5%). Conduct a hypothesis test to determine if Providence College satisfies Prong I of Title IX.

Hypothesis Test
1. H_0:
2. H_A:
3. Test statistic and its value:
4. *p*-value:
5. Your statistical conclusion:
6. Your interpretation:

Does the University of Vermont (UVM) satisfy Prong I of Title IX? They dropped football and now have only 7 sports for men, which they needed to do as their undergraduate population is only 43.7% male and therefore 56.3% female. The observed frequency of athletes at UVM is 185 for the men and 216 for the women.

Perform a hypothesis test to determine if UVM satisfies Prong 2 of Title IX.

Hypothesis Test
1. H_0:
2. H_A:
3. Test statistic and its value:
4. *p*-value:
5. Your statistical conclusion:
6. Your interpretation:

Does the University of Hartford satisfy Prong I of Title IX? 51% of their undergraduates are female, and 49% are male. But, even without football and wrestling (two sports they have dropped), they have 146 female athletes and 168 male athletes. Do a hypothesis test (list H_O and H_A) to find your chi-squared test statistic and your p-value. Then type in your statistical interpretation and your conclusion.

Hypothesis Test
1. H_O:
2. H_A:
3. Test statistic and its value:
4. *p*-value:
5. Your statistical conclusion:
6. Your interpretation:

Trending now in sports

PREVIEW: ANOTHER GRAPH FOR INVESTIGATING DATA OVER TIME

A **time-series plot** graphs data which has been collected over time, thus the name time-series plot. For example, a rookie running back in the NFL may start out his rookie season with huge numbers. But after four exhibition games, then a sixteen game regular season with only one bye week, then potential playoff games, against bigger, faster and stronger players than what he faced in college, the extra strain of the NFL competition will start to wear him down. If we plotted his yards per game for his rookie year over his 16 regular season games, we would probably see a decline in the number of yards per game. We could do the same with his yards per carry.

This plot of time-series data will reveal trends to opposing teams and his own coaches. These are trends over time. So the x axis is always time, whether it be years or weeks or days or hours. The y axis is always the data, in our example yards per carry or yards per game. You will put a dot on the graph representing each data point and then at the end connect the dots with a line to see the trend.

For example, let's examine Wayne Gretzky's incredible career. He scored 894 goals in 1487 games, better than one goal every two games for 21 years! Enter your numbers to analyze into C1. In this case, I copied Gretzky's career goals per season and pasted them into C1. Then go to "Graph", "Time Series", and keep it "Simple". Double click on C1 to move it into the series box, then click labels. For labels I added titles and subtitles to put the numbers into context, subtitle 2 became my name, and in the footnote enter your interpretation of the graph. Namely, what do you see happening over the course of Gretzky's career?

Then click OK.

For now, this line just gives us the trend. It tells us whether our data is rising or falling, or even flat. In a few weeks, when we add a second quantitative variable, this line will become our regression line!

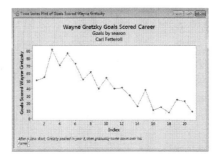

Another use for a time series plot is a comparison. For instance, David Ortiz just retired and the Red Sox have yet to replace his bat. There are two free agents with a history of power play for the Blue Jays: Edwin Elpidio Encarnación and Jose Bautista. Encarnación has 310 home runs in 12 seasons, while Bautista has 308 home runs in 13 season. So while their careers are similar, we can use a time series plot to look at the recent trends, in this case, for power.

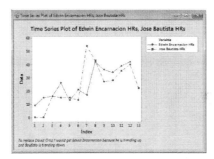

So Encarnación is a better choice for the Red Sox to replace Ortiz because he is trending up the last two seasons while Bautista had a large drop-off this year.

TIME-SERIES PLOT EXERCISE

Sidney Crosby is one of the premier players in the NHL, and was the team Canada captain for the Sochi Olympics. Look up Sidney Crosby's stats. Create a time-series plot of his goals scored over the course of his career. Label the graph. Then use the footnote of the graph for your interpretation.

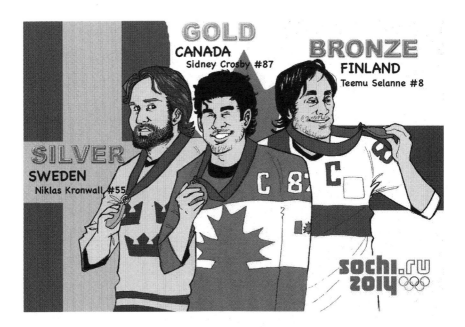

Gaze into your crystal ball

PREVIEW: PREDICTING THE FUTURE WITH THE FINAL GRAPH, A SCATTERPLOT

Making predictions about future performance

After you graduate from Springfield College with your minor in Sports Analytics, you may get a job as general manager of the Boston Red Sox. Your first task is to upgrade the team's offense. You are offered Edwin Encarnación and you wonder how he would fit in with your line-up in Boston. So you plug his stats into your team's data, and you predict Encarnación's performance for the next season with the Boston Red Sox. That prediction you made? That is regression. Simple, right? Now let's see how regression works.

To start with, we need two quantitative variables, like ERA and strikeouts, or OBP and OPS. Go to baseball-reference.com and find the 2015 stats for the Boston Red Sox.

I mentioned OBP and OPS above, so let's use those. OBP is On Base Percentage, and OPS is On Base Percentage plus Slugging. I want to see if I have a **linear relationship** between those two variables before I use them for making predictions. A linear relationship means it can be described with a line, that is when one value goes up or down, the other goes up or down. They're tied together. If the relationship is linear, we can use one variable to predict the other.

There are two ways to verify if there is a linear relationship. One is with a graph, the other is mathematically. We will do both. And just like categorical and quantitative variables, we will start with the graph first. But this time the graph is a new one for us. It is called a **scatterplot**. To graph a scatterplot I copied the columns labeled OBP and OPS into Minitab (see below).

Then click the "stat" link, choose "regression" and then click on "fitted line plot". The fitted line plot is the scatterplot. On the x axis you will put the **independent variable**, also known as the explanatory variable, and on the y axis you will put the **dependent variable**, or the response variable. Minitab asks for y first, which seems backwards, but just make sure you click the y first because of that.

So which variable is y? It is the result. Like comparing 5K time and miles per week. Your 5K race result *depends* on how many training miles you ran. So the 5K time is the y. Or hours studying and your grade on the next exam. Your grade is the result and will depend on how many hours you study. So your grade is the y. How about OBP and OPS? OBP is a part of OPS, so OPS depends on your OPB. OPS will be our y and then OBP will be our x.

Note: There are times when one variable is not the result of the other, like comparing ERA and WHIP. You could easily argue that ERA depends on WHIP, but the same argument holds true for WHIP depending on ERA. In that case it does not matter which is y or x. In class, we will put the first variable in as y and the second as x so we can all calculate the regression equation the same, but you would not be wrong doing it the other way.

So double click on OPS to move it to y and OBP to x. Click OK and the scatterplot will look like this:

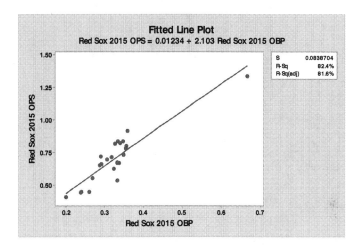

Remember, we are looking for a relationship between OBP and OPS. If we see a pattern in the dots on our scatterplot, there is a relationship.

In the graph above, we can see the dots clustering. We do see a pattern. As Minitab shows we could put a line through the data points. So we have a linear relationship! Yes, a relationship does exist between OBP and OPS.

The closer the dots are together, the stronger the relationship. If the dots formed a line, it would be a very strong relationship. Our dots are close, but not that close. We would say we have a strong, linear relationship.

One more aspect of this new relationship we have formed is direction. If the line is going up (from left to right) as above, it is **positive**. If the line is going down from left to right, it is **negative**. So we have a strong, positive, linear relationship.

Note: Positive is not good and negative is not bad. Positive in graphing means that as one value goes up, the other value goes up. Or as one value goes down, the other value goes down. They go in the same direction. That makes sense. In baseball, when a player gets hot and his OBP goes up, we would expect his OPS to go up as well. Also, when he slumps and his OPB goes down, we would expect his OPS to go down too. So we would have guessed before we started that, if a relationship existed, it would have been positive.

An example of a negative relationship would be as one value goes up, the other goes down! They go in different directions. Like training miles per week and 5K time. If you can squeeze in more training, the miles per week increases. Increasing training miles hopefully leads to *decreased* 5K times! So as one goes up, the other goes down. We call that a negative relationship. Not in a bad (or negative) way though.

Remember, our ultimate goal was to predict a player's OPS when he is plugged into the Red Sox lineup, using his OBP. We can make that prediction IF we have a linear relationship. Based on the graph we generated on Minitab, we decided we do have a relationship, specifically a strong, positive, linear relationship.

But, if the dots were further apart we might not be certain of a linear relationship. We need a mathematical way to check, a method that is black and white and leaves us with no doubt about whether a relationship exists. That tool is called **correlation**. Look at the root of the word correlation and you see relation, so it is a measure of a relationship of two variables.

To calculate correlation on Minitab we have two options. On the graph Minitab created, it gave us r^2 (R-Sq in the upper right). In statistics it is standard to use r for correlation or, more specifically, the Pearson correlation coefficient. We could just take the square root of that value and that would give us r. Or we can go to stat, basic stat, correlation, and put the two variables in the box, and say OK. It will give us the **Pearson correlation value**. (Karl Pearson was a big deal in mathematical statistics. He developed the hypothesis testing we do in this course. The official name for the chi squared test is the Pearson chi squared test. He is also generally credited for the histogram.)

But what does this number mean? r can take on any value between -1 and 1, inclusive. The closer r is to -1 or 1, the stronger the relationship. The closer r is to 0, the weaker the relationship. In our example, we have an r of 0.91, which is pretty close to 1, so we have a strong relationship. Also, r is positive, so we have a strong, positive relationship. Which agrees with our graph.

Correlation OBP, OPS

```
Pearson correlation of OBP and OPS = 0908
P-Value = 0.000
```

Basically r measures the strength and direction of a linear relationship. We can only use r when the relationship is linear. And r will match the slope. If r is positive, the slope is positive, and the regression line goes up from left to right. If r is negative, the slope is negative, and the regression line goes down from left to right.

Which is stronger, an r of 0.89 or an r of 0.65? 0.89 is because it is closer to 1. How about an r of -0.77 or an r of -0.92? -0.92 because it is closer to -1. The closer you are to -1 or 1, the stronger the relationship. How about -0.71 or 0.71? Neither! They are equally strong because they are the same distance from the endpoints.

Now r^2. r^2 is another measure of the strength of our model. When you loaded the data into Minitab for OBP and OPS, and then ran regression, you created a model. That model is what we will use to predict OPS from OBP now that we have determined through the graph, and confirmed with r, that a linear relationship exists.

In this case r^2 gives us the percent of the variation in OPS that can be explained by the batter's OPB. In general, it is always the percent of the variation in y that can be explained by x. Those y and x values will change with each problem we do. We can find r^2 in two ways. First the easy way, it is on your scatterplot! Of the two choices, use R-Sq, not R-Sq(adj), as R-Sq represents the data better. Second, R2 is literally r squared (r^2). So yes, you can do r times r on your phone.

On your graph you also have at the top the actual regression equation. In our case, it is:

Red Sox 2015 OPS = 0.01234 + 2.103 Red Sox 2015 OBP

What does this equation mean? Remember, we are trying to predict OPS for a new Red Sox player. Well, that is the left side! Red Sox 2015 OPS =. That is where our prediction will show up. The next is 0.01234. That is the y intercept, which we have to be very careful with in regression, because it is either outside of our data's range or just plain meaningless. For instance, if a player has an OBP of 0, the y intercept says he will have a predicted OPS of 0.01234! Impossible. Look at any player on the Red Sox, and you will see plenty of pitchers with an OBP and slugging of 0, so when you add them together, 0 + 0 = 0, right? Not 0.01234! Because of this, we rarely use the y intercept with regression, and if we do, we do it with caution. We won't see it again in this course.

But the next number we will see again. 2.103 is the slope. Notice the slope is positive, as it should be since the regression line goes up from left to right. And the positive slope agrees with r, in terms of both being positive. But the slope also has a deeper meaning, it tells us how much our y will change based on a unit change in x. Huh? Well, in our case, if a player's OBP goes up by 1, his OPS goes up by 2.103! So a hitting coach who improves a player's OPB, will affect his OPS by more than twice the improvement for OBP.

Now we can try to predict a player's OPS for an OBP of 0.360. On Minitab you will do stat, regression, regression, fit the regression model. Then OPS will be your response, and OBP will be your predictor. Then say OK. The next time you do stat, regression, regression, the predict option will be available to you. Use that and type in 0.360, then OK. Your model will give you your prediction for a player's OPS, using any OBP numbers you type in! On Minitab it is labelled as "fit" in the output.

```
Regression Analysis: OPS versus OBP

Prediction for OPS

Regression Equation
OPS = 0.0123 + 2.103 OBP

Variable    Setting
OBP            0.36

Fit         SE Fit         95% CI                  95% PI
0.76949     0.018596       (0.73092, 080805)       (0.59133, 094765)
```

By hand you will plug 0.360 into your regression equation. Simply multiply 0.360 times 2.103. Take that answer and add 0.01234. You get 0.76942. They agree! Now, are you comfortable making that prediction? Look at your Red Sox data for 2015. 0.360 is in the range of your data, at the very top, but in the range. So yes, you are comfortable. If I ask you to make a prediction for an OBP of 0.370, you could, but you would be uncomfortable. Our model is based on OBP's up to 0.360, so we do not know how the model would react to data points that high. So as long a I give you values within the range of your data, you are comfortable. Outside of that, you are not comfortable.

When you look at the Red Sox data, one of the players had an OBP of 0.360, so why am I asking you to predict that? We already know his OPS. First, I could have asked you to predict OPS for an OBP of 0.359, and no Red Sox had an OBP of 0.359. Second, now I can measure how well the 0.360 player did. Our model says he should have an OPS of 0.769, he actually had an OPS of 0.913, much higher! He outperformed our model, so he over performed in 2015.

The difference between his actual performance (0.913) and his predicted (0.769) is called his **residual**. So he had a positive residual of 0.144, and anytime a player has a positive residual they over-perform, and a negative residual means they underperform.

Note: A note on residuals. Notice on the scatterplot earlier, the regression line goes through the data points. The difference between any data point and the line is the residual. See if you can guess which dot belongs to our 0.360 hitter based on that. It was not arbitrary how Minitab placed that line. Minitab calculated the residuals for each possible line, squared the residuals, and added them up. The line that had the smallest total of squared residuals was rewarded with showing up on our graph and the honor of being called the regression line.

Which is the best predictor?

PREVIEW: FIND THE BEST WAY TO PREDICT THE FUTURE

In the Chapter 17 we learned about residuals. The residual is the difference between the actual values and the predicted values. The predicted values came from the regression model (the line) Minitab drew for us. On the graph we could see residuals. We could see how far away each point was from the regression line. Some points were above the line meaning their residuals were positive. Some below and had negative residuals. And some points were on the line. They had a residual of 0.

Remember how I said a test statistic turns an entire set of data into a single number? We can do the same things with residuals. We can gather all those residuals together into a single number. It's called the **root mean square error** (RMSE). The RMSE (also called the root mean square deviation, RMSD) is the average of all the residuals. That is, it measures the difference — the error or deviation — of the observed values from the predicted values (the values represented by the line).

To do it by hand you would subtract each actual value from the predicted value. Square them to get rid of the negatives. Add them all up. Divide by the number of values which gives the average (mean) of how far the data deviates from predicted. Then take the square root. Here's what the formula looks like:

$$RMSE = \sqrt{\frac{\sum_{i=1}^{n}(X_{obs,i} - X_{model,i})^2}{n}}$$

Where X_{obs} is the actual (observed) values and X_{model} is the predicted values from your model.

Fortunately Minitab will calculate RMSE for us.

When we did the scatterplot in the regression chapter (Chapter 17), we used the fitted line plot. There was one value on the graph summary we did not use in regression, and that was S. Well, S is short for RMSE! So when we run the regression model and generate our graph, Minitab calculates RMSE for us, and then displays the result as S.

Let's try an example. When I look at the stats for MLB hitters in the American League in 2016, I notice the final column is WAR (wins above replacement). It is a relatively new stat that's supposed to sum up a player's total contributions to their team. Supposedly it's the number of additional wins a player contributes above what's expected from a replacement-level player.

	C1	C2	C3	C4	C5
	AVG	OBP	SLG	OPS	WAR
2	0.325	0.388	0.457	0.845	5.2
3	0.323	0.439	0.578	1.018	9.2
4	0.320	0.406	0.629	1.035	4.2
5	0.320	0.371	0.401	0.773	1.5
6	0.319	0.363	0.461	0.824	5.2
7	0.319	0.359	0.560	0.919	7.8
8	0.312	0.387	0.545	0.932	4.2
9	0.307	0.364	0.453	0.816	2.9
10	0.305	0.358	0.565	0.922	6.4
11	0.305	0.360	0.454	0.813	2.7
12	0.302	0.352	0.526	0.877	6.1
13	0.297	0.407	0.578	0.985	6.9
14	0.295	0.353	0.405	0.757	1.9
15	0.295	0.351	0.482	0.833	1.5
16	0.295	0.337	0.520	0.858	1.7

I am curious which of the old stats I am familiar with are the better predictor of WAR. I loaded Average, OBP, Slugging, OPS and WAR into Minitab (C1 through C5).

Remember that regression is just a fancy name for prediction. So I will do the exact same steps that I did in the regression section. I will graph each predictor separately. So the first graph (on the next page) shows Average as a predictor for WAR and the resulting S (RMSE, that is) = 1.63. S shows up on the worksheet AND the graph, so it is easy to find. But I did stat, regression, fitted line plot. I put WAR as the response, and AVG as the predictor. And then OK.

I have to do this four times, because I have four predictors. So I have to find S for each one. Each time I enter the information for the graph, I do not change WAR. WAR is my response for all four graphs. But I change my predictor each time.

Here is the graph for AVG and WAR.

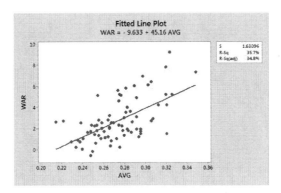

And the graph for OBP and WAR:

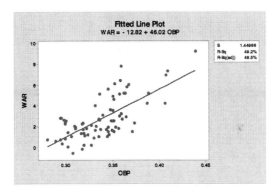

And SLG and WAR:

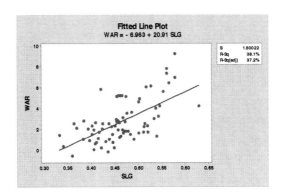

And OPS and WAR:

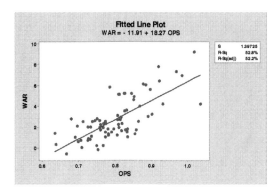

To summarize:

Predictor	RMSE
AVG	1.631
OBP	1.445
SLG	1.600
OPS	1.397

The RMSE is much higher for AVG. This tells us that the residuals are much larger for AVG than any of the other three predictors. Remember the E in RMSE stands for error. So a large RMSE means a large error. We are looking for the best predictor. It is certainly not average.

The best predictor will always be the lowest RMSE. We want to minimize error, and the lower the number, the lower the error. So OPS is the best predictor of WAR (at least in the AL in 2016) because it has the lowest RMSE.

CHAPTER 19

Borda count

PREVIEW: ALL MVP AWARDS, HEISMAN TROPHIES, TOP 25 POLLS USE THIS METHOD

The Borda count is an election method in which voters rank candidates in order of preference. The Borda count determines a single winner of an election by giving each candidate a certain number of points corresponding to the position in which he or she is ranked by each voter. Once all votes have been counted the candidate with the most points is the winner. Because it sometimes elects broadly acceptable candidates, rather than those preferred by the majority, the Borda count is often described as a consensus-based electoral system, rather than a majoritarian one.

The Borda count was developed independently several times, but is named for the 18th-century French mathematician and political scientist Jean-Charles de Borda, who devised the system in 1770.

The Borda count, and points-based systems similar to it, are often used to determine awards in competitions. The Borda count is a popular method for determining awards for sports in the United States, and is used in determining the Most Valuable Player and Cy Young Award Winner in Major League Baseball, by the Associated Press and United

Press International to rank teams in NCAA sports, to determine the winner of the Heisman Trophy, and so on.

So let's look at college football for example. The Associated Press poll is a poll of football writers across the country. Each eligible voter, and there are 60 of them, receive a ballot and each Sunday they fill out who they think are the top 25 teams in order of preference. They then submit the ballot to the AP who calculates the points earned by every team using the above mentioned Borda Count method. For each ballot, the 25 ranked team gets one point, the 24th ranked team gets two points, and the 23rd ranked team gets three points, up to the top ranked team who gets 25 points. The AP then adds up all the points and publishes their rankings every Monday. Here's the AP poll for 2010.

The Associated Press Top 25 Poll
Monday October 11, 2010

Rank	Team	Record	Votes	Previous
1	Ohio State (34)	6-0	1,453	2
2	Oregon (15)	6-0	1,427	3
3	Boise State (8)	5-0	1,395	4
4	TCU (1)	6-0	1,304	5
5	Nebraska	5-0	1,236	7
6	Oklahoma (2)	5-0	1,225	6
7	Auburn	6-0	1,104	8
8	Alabama	5-1	1,021	1
9	LSU	6-0	999	12
10	South Carolina	4-1	978	19
11	Utah	5-0	926	10
12	Arkansas	4-1	813	11
13	Michigan State	6-0	806	17
14	Stanford	5-1	732	16
15	Iowa	4-1	648	15
16	Florida State	5-1	547	23
17	Arizona	4-1	472	9
18	Wisconsin	5-1	410	20
19	Nevada	6-0	376	21
20	Oklahoma State	5-0	348	22
21	Missouri	5-0	298	24
22	Florida	4-2	209	14
23	Air Force	5-1	187	25
24	Oregon State	3-2	186	NR
25	West Virginia	4-1	141	NR

SPRINGFIELD COLLEGE
Sports Statistics Exam 2 Spring 2015

Springfield College Field Hockey had a good 2014, going 9-8 in the regular season and reaching the NEWMAC Quarterfinals (losing 1-0 to Babson). So let's look at their statistics to see if there is a relationship between the shots they take (SH) and the points they accumulate (PT).

2014 Springfield College Field Hockey

#	Name	Yr	Pos	GMS	STS	GL	AS	PT	SH
2	Tori Stollmer	So	F	7	-	1	1	3	5
3	Jillian Macdonald	So	D	14	5	0	0	0	0
4	Ann Mahoney	Fr	M	17	7	1	1	3	10
5	Kristina Krull	Jr	F	18	15	6	4	16	46
6	Sarah Ducharme	Sr	D	17	8	4	0	8	10
7	Olivia Cabral	Jr	F	18	18	12	6	30	88
8	Jess Lawson	Sr	D	17	13	0	0	0	2
9	Linnea Quist	Sr	M	18	18	10	4	24	50
10	Tina DeGirolamo	Sr	D	17	12	0	0	0	0
11	Paige Laperle	Fr	M	14	6	0	1	1	2
12	Maggie Kennedy	Fr	F	16	3	4	2	10	36
14	Sarah Kelly	Jr	F	5	-	0	0	0	2
16	Dylan Chisholm	Jr	F	1	-	0	0	0	0
17	Jamie Raccosta	So	D	18	18	1	3	5	9
18	Libby Hart	Jr	M	18	18	0	4	4	7
20	Mikayla Wysocki	So	D	14	4	0	0	0	0
21	Danielle Sweet	Fr	M	18	18	0	0	0	1
22	Katie Gill	Jr	M	18	17	6	4	16	34
23	Emily Greenwell	Fr	M	3	-	0	0	0	1
24	Megan Rooney	Jr	F	2	-	1	0	2	1
31	Timarie Villa	Jr	GK	16	16	0	0	0	0
33	Juliana Barrasso	So	GK	4	2	0	0	0	0

1. Draw a scatterplot and print it out.
2. Based on the scatterplot, is there a relationship?
3. Why?
4. Calculate the correlation r
5. Interpret r
6. Calculate r^2
7. Interpret r^2
8. Based on r and r^2, is there a relationship?
9. What is the regression equation?
10. If a player has 60 shots, how many points would you predict for her?
11. Are you comfortable with this prediction and why?
12. If a player has 100 shots, how many points would you predict for her?
13. Are you comfortable with this prediction and why?
14. Kristina Krull had 46 shots, how many points would you predict for her?
15. What is her residual?
16. Interpret her residual.

Look up Dustin Pedroia's career stats. Create a time series plot for his OPS (on base percentage plus slugging) over the course of his career. Label the graph. Then comment on the graph.

Does AIC satisfy Prong I of Title IX? They have an undergraduate population that is 42% male and 58% female, yet their observed frequency for women athletes is 175 and for men 341.

Do a hypothesis test (list H_0 and H_A) to calculate your chi-squared test statistic and your *p*-value. Then type your statistical conclusion and your interpretation.

We saw on the first day of Sports Stats that a QB Rating is made up of Touchdowns, Completions, Interceptions and Yards. But which one is the best predictor of an NFL QB's rating? Use RMSE (called S on Minitab) to decide which of the four stats is the best predictor of QB rating and why.

NFL Player Passing Statistics - 2014

Rk	Player	Team	Pos	Cmp	Att	Pct	Att/G	Yds	Avg	Yds/G	TD	Int	Rate
1	Drew Brees	NO	QB	456	659	69.2	41.2	4,952	7.5	309.5	33	17	97.0
1	Ben Roethlisberger	PIT	QB	408	608	67.1	38.0	4,952	8.1	309.5	32	9	103.3
3	Andrew Luck	IND	QB	380	616	61.7	38.5	4,761	7.7	297.6	40	16	96.5
4	Peyton Manning	DEN	QB	395	597	66.2	37.3	4,727	7.9	295.4	39	15	101.5
5	Matt Ryan	ATL	QB	415	628	66.1	39.2	4,694	7.5	293.4	28	14	93.9
6	Eli Manning	NYG	QB	379	601	63.1	37.6	4,410	7.3	275.6	30	14	92.1
7	Aaron Rodgers	GB	QB	341	520	65.6	32.5	4,381	8.4	273.8	38	5	112.2
8	Philip Rivers	SD	QB	379	570	66.5	35.6	4,286	7.5	267.9	31	18	93.8
9	Matthew Stafford	DET	QB	363	602	60.3	37.6	4,257	7.1	266.1	22	12	85.7
10	Tom Brady	NE	QB	373	582	64.1	36.4	4,109	7.1	256.8	33	9	97.4
11	Ryan Tannehill	MIA	QB	392	590	66.4	36.9	4,045	6.9	252.8	27	12	92.8
12	Joe Flacco	BAL	QB	344	554	62.1	34.6	3,986	7.2	249.1	27	12	91.0
13	Jay Cutler	CHI	QB	370	561	66.0	37.4	3,812	6.8	254.1	28	18	88.6
14	Tony Romo	DAL	QB	304	435	69.9	29.0	3,705	8.5	247.0	34	9	113.2
15	Russell Wilson	SEA	QB	285	452	63.1	28.2	3,475	7.7	217.2	20	7	95.0
16	Andy Dalton	CIN	QB	309	481	64.2	30.1	3,398	7.1	212.4	19	17	83.5
17	Colin Kaepernick	SF	QB	289	478	60.5	29.9	3,369	7.0	210.6	19	10	86.4
18	Brian Hoyer	CLE	QB	242	438	55.3	31.3	3,326	7.6	237.6	12	13	76.5
19	Derek Carr	OAK	QB	348	599	58.1	37.4	3,270	5.5	204.4	21	12	76.6
20	Alex Smith	KC	QB	303	464	65.3	30.9	3,265	7.0	217.7	18	6	93.4
21	Cam Newton	CAR	QB	262	448	58.5	32.0	3,127	7.0	223.4	18	12	82.1
22	Kyle Orton	BUF	QB	287	447	64.2	37.2	3,018	6.8	251.5	18	10	87.8
23	Teddy Bridgewater	MIN	QB	259	402	64.4	30.9	2,919	7.3	224.5	14	12	85.2
24	Blake Bortles	JAC	QB	280	475	58.9	33.9	2,908	6.1	207.7	11	17	69.5
25	Geno Smith	NYJ	QB	219	367	59.7	26.2	2,525	6.9	180.4	13	13	77.5

SPRINGFIELD COLLEGE
Sports Statistics 🏈 Exam 2 Fall 2015

The Springfield College Pride football team is currently (through November 7, 2015) ranked seventh in the country in rushing offense with just over 300 yards per game. We will use regression to see if there is a relationship between rushing attempts (rush) and yards gained (yds) for the 2015 Springfield College team as shown in the table below.

1. Draw a scatterplot and print it out.
2. Based on the scatterplot, is there a relationship between rushing attempts and yards gained?
3. Why do you say that?
4. Calculate the correlation r.
5. Interpret r.
6. Calculate r^2.
7. Interpret r^2.
8. Based on r and r^2, is there a relationship between rushing attempts and yards gained?
9. What is the regression equation?
10. If an SC player had 75 rushing attempts, how many yards would you predict for him?
11. Are you comfortable with this prediction and why?
12. Sam Benger ('18) of Carnegie-Mellon University leads the country in DIII rushing yards with 1660 yards on 242 rushes. If he played for SC and he rushed the ball here 242 times, how many yards would you predict for him?
13. Are you comfortable with this prediction and why?

14. Billy DeLay (SC '17) had 6 rushes this year before his concussion, how many yards would you predict for him?

15. What is his residual?

16. Interpret his residual.

Look up David Ortiz's stats at ESPN.com. Copy his career HR's into Minitab. Then create a time series plot for his home runs hit. Label the graph, including your name, then comment on the graph.

Which is the better predictor of a MLB player's power? Use runs, hits, doubles and triples to see which is the better predictor of home runs.

Go to MLB.com. Click on stats. On that screen, copy R, H, 2B, 3B and HR into Minitab and then use RMSE (S on Minitab) to decide which is the better predictor of home runs and why.

Does the University of Notre Dame satisfy Prong I of Title IX? They have an undergraduate population that is 52% male and 48% female. Their observed frequency for male athletes is 421 and 316 for the women.

Do a hypothesis test (list H_0 and H_A) to calculate your test statistic and your p-value. Then enter your statistical conclusion and your interpretation.

SPRINGFIELD COLLEGE
Sports Statistics Exam 2 Spring 2016

The 2015/16 Golden State Warriors are on pace to break the all-time NBA record for wins in a season, set by the 95/96 Chicago Bulls, who finished 72-10. So we will focus on their accomplishments this year. Search "Golden State Warriors Stats" and click on the ESPN link. Copy five columns into Minitab: PPG, RPG, APG, SPG, and PER. Those are points, rebounds, steals, assists per game and player efficiency rating.

In question one, we will examine the relationship between PPG and PER only. Save the other columns for question 2.

1. Draw a scatterplot and print it out.
2. Based on the scatterplot, is there a relationship between PPG and PER for the Golden State Warriors?
3. Why do you say that?
4. Calculate the correlation r.
5. Interpret r.
6. Calculate r^2.
7. Interpret r^2.
8. Based on r and r^2, is there a relationship between PPG and PER for the Golden State Warriors?
9. What is the regression equation?
10. Draymond Green has 13.8 PPG, what PER would you predict for him?
11. Are you comfortable with this prediction and why?
12. In Michael Jordan's best season, he averaged 37.1 PPG. If he did that this season for the Warriors, what would you predict for his PER?
13. Are you comfortable with this prediction and why?

14. Klay Thompson is the number two scorer on this team, with 22.6 PPG. What would you predict for his PER?

15. What is his residual?

16. Interpret his residual.

Let's take a further look at the new stat for basketball: Player Efficiency Rating (PER). To put PER in context, only one NBA player is over 30 this season, and that is Stephen Curry. But what old stat is the best predictor of an NBA player's PER? Use points (PPG), rebounds (RPG), assists (APG) and steals (SPG) to see which is the better predictor of PER.

To see which is the better predictor use RMSE (S on Minitab) to analyze the data, then list all four S values, then interpret the values to determine which is the better predictor of PER.

Search "Stephen Curry career stats" and click on basketball-reference.com. Copy his total points (the last column – PTS) into Minitab and create a time-series graph of his points scored over his career (include this year). Label the graph, including your name, then comment on the graph.

Only one school has teams in both the men's and women's final four ... the Orangemen! But do they treat their programs equally? We will do a chi-squared test to see if they satisfy Prong I of Title IX.

They have an undergraduate population that is 45% male and 55% female. Their observed frequency for male athletes is 263 and 275 for the women.

Do a hypothesis test (list H_0 and H_A) to calculate your test statistic and your *p*-value. Then enter your statistical conclusion and your interpretation.

SPRINGFIELD COLLEGE
Sports Statistics Exam 2 Fall 2016

The University of Alabama won the national championship in NCAA Division I football in 2016, their 4th national title in the last seven years. Because their campus is mostly women, they have a hard time meeting Title IX requirements. So they have dropped the following men's sports: volleyball, track and field, wrestling, gymnastics, and soccer! Is that enough to put Alabama in compliance?

So answer the question: Does the University of Alabama satisfy Prong I of Title IX? They have an undergraduate proportion that is 56% female and 44% male. Their observed frequency for female athletes is 263 and for male athletes is 238.

Do a hypothesis test (list H_o and H_A) using the chi-squared goodness of fit test to find your test statistic and p-value. Then state your statistical conclusion, and based on the statistics, your interpretation about Alabama.

We saw at the beginning of the year that QB Rating was calculated partly from a quarterback's completions and yards. Let's see if there are other factors that might be better predictors. We will use Comp (completions), ATT (attempts), PCT (completion percentage) and YDS (passing yards) to see which the better predictor of QB Rating is.

Google "NFL Stats". Click on the NFL.com link. Under Passing Yards, click on the "complete list" link. Copy the entire columns: Comp, Att, Pct, Yds and Rate into Minitab. Then use RMSE (S on Minitab) to decide which is the better predictor of QB Rating (Rate).

Using the same NFL data, we will investigate to determine if there is a relationship between Pct (completion percentage) and Rate (QB rating).

1. Draw a scatterplot and print it out.
2. Based on the scatterplot, is there a relationship between completion percentage and QB rating for NFL quarterbacks?
3. Describe the relationship
4. Calculate the correlation r
5. Interpret r
6. Calculate r^2
7. Interpret r^2
8. Based on r and r^2, describe the relationship between Pct and Rate for NFL quarterbacks
9. What is the regression equation?
10. Jimmy Garoppolo had a 70% completion percentage (Pct). What would you predict for his QB Rating (Rate)?
11. Are you comfortable with this prediction and why?
12. In 2015, Luke McCown of New Orleans led the NFL with a completion percentage of 82.1%. What would you predict for his QB rating this year?
13. Are you comfortable with this prediction and why?
14. Tom Brady is number one in 2016 with a 73.1% completion percentage. What would you predict for his QB Rating (Rate)?
15. What is his residual?
16. Interpret his residual.

While the Red Sox are deciding who they can bring in to replace David Ortiz, a possible answer is already wearing a Red Sox uniform (albeit a XXXL). That is Pablo Sandoval, a non-contributor since his arrival from San Francisco. Let's check his OPS trend to see if he would be a candidate to replace Ortiz

Google "Pablo Sandoval stats" then click on the ESPN link. Copy his OPS column into Minitab and create a time-series graph of his OPS for his career. Label the graph, including your name, then comment on the graph.

PART 3
Statistical Inference

Confidence intervals for 1 proportion

PREVIEW: A NEW TEST. WE'LL USE IT WITH 1 PROPORTION

The last section of this course deals with statistical inference. Inference is when we make an estimate of a population based on a sample. We *infer* — deduce or conclude. Like Sherlock Holmes. We take what we know to make an educated guess at something we want to know. **Statistical inference** is when we use statistical tests to make that deduction. The two tests we will use are the confidence interval and the hypothesis test. You have already seen the hypothesis test in action with chi-squared when we compared observed frequencies with expected frequencies. So we will begin this last section with confidence intervals.

A confidence interval gives us an **interval** or range in which our estimate of the population could fall. That interval is associated with a level of **confidence**, usually 90, 95, or 99%. But it could be anything, 88%, 92%, etc. 90%, 95%, and 99% are just the most common. Basically, the more important the question, the more confident you want to be, and, the higher the confidence level you want. In sports stats, our questions are important to us, but not life and death, like the FDA approving a drug that could have fatal side effects. So the FDA would use 99% and up, we will stick with 95%. So whenever we do confidence intervals, we will always do a 95% confidence interval.

Let's use David Ortiz for example. He played for the Minnesota Twins for six years before coming to Boston and becoming an integral part of the Boston Red Sox World Series win in 2004. But could we have anticipated the success Ortiz had in his first five seasons with the Red Sox? Yes! How can I answer that so definitely? By using a confidence interval. Here is how it works.

Overall in Ortiz's six years with the Twins, he had 393 hits in 1477 at bats, for a batting average of 0.266. Since the 1477 at bats with the Twins is a healthy sample of at bats, I am going to use that sample to estimate what Ortiz will hit in 2003 with the Red Sox. But I can't predict a single number for his average, because batting average is a variable. So I will use the 0.266 that he averaged for six seasons and then create an interval by adding and subtracting the same amount from 0.266. The Red Sox did the same thing before he became part of the team, they created a window, or an interval, that they thought his batting average would fall into.

That interval was (0.243688, 0.289406), or 0.266 +/- 0.022. The 0.266 is called a point estimate. It's an estimate of what he will bat this year based on his history with the Twins. Then they added and subtracted 0.22 from 0.266 to get the actual interval. The 0.22 is called the margin of error.

The margin of error is,

$$z^* \sqrt{\frac{p-q}{N}}$$

where z^* is the critical z value, p is the sample proportion, in this case 0.266. q is 1-p, of 1 - 0.266, or 0.734. N is the sample size, in this case his 1477 at bats.

So a confidence interval is found by taking the point estimate then adding and subtracting the margin of error.

So the Sox thought he would hit between 0.244 and 0.289 for them in 2003. He hit 0.288 that year! So the Red Sox were right on. They created a 95% confidence interval for what they thought Ortiz would hit, and then Ortiz hit in that interval. As a bonus for Red Sox management, he even hit at the upper end of the interval.

Here is how you can create the same interval on Minitab. Go to stats, basic stat, and then choose one proportion. Why proportion? Well, we saw earlier that Ortiz with the Twins had 393 hits in 1477 at bats, or as a fraction, was 393/1477. Anytime you see a fraction, or a decimal (0.266

when you divide 393 divided by 1477) or a percent, those are examples of proportions. So that is your clue on Minitab to use proportion.

Minitab defaults to one or more samples, each in a column. No, we have summarized data. We have "summarized" David Ortiz's six years with the Twins into 393/1477. So choose summarized data. The number of events is the top number of our fraction, namely his hits. And the number of trials is the bottom number, namely his at bats. So enter 393 and 1477 into each box respectively. Then say OK.

```
— — — — — 9/5/2016 8:15:08 AM — — — — —

Welcome to Minitab, press F1 for help.

Test and CI for One Proportion

Sample    X      N     Sample p           95% CI
1        393   1477   0.266080    (0.243688, 0.289406)
```

The results are (0.243688, 0.289406), a p of 0.266 and n of 1477. So n we know, p is his batting average, and the values in the parentheses represent our 95% confidence interval. So, the Red Sox were 95% confident that Ortiz would hit between 0.244 and 0.289 in 2003, and he did, hitting 0.288.

So, let's look ahead five years. From 2003 through 2007 Ortiz batted 0.302, well above the confidence interval values the Red Sox had predicted for him. Since he fell in the interval in 2003, this is further evidence of his use of performance enhancing drugs in 2004 and beyond. Then when testing began, he fell to 0.264 in 2008, then 0.238 in 2009!

We will look to see if his drop in 2008 was just a matter of random variation, since batting average is a variable, or if it was old age catching up to a local icon. From 2003 through 2007, he had 826 hits, so we enter that for "number of events". Then for "number of trials" enter his total at bats over that period of 2738. Again we will stay with a 0.95 for our confidence level, and calculate.

```
— — — — — 9/5/2016 8:54:18 AM — — — — —

Welcome to Minitab, press F1 for help.
```

Test and CI for One Proportion

```
Sample    X      N     Sample p         95% CI
1        826   2738    0.301680    (0.284525, 0.319260)
```

Assuming Ortiz was the same player in 2008 as he was in the five previous seasons, that calculation tells us he would have hit between 0.285 and 0.319 in 2008. He hit 0.264, well below the interval. Now we have more evidence that there is something going on here, it is not just random variation.

One more. On May 5, 2010, Ortiz was 10 for 67 and batting 0.149! NESN was running a poll asking if he should be benched or released. Talk radio was killing him. And his manager stood by him. Francona did not bow to pressure because he believed in Ortiz. He left David in the line-up.

```
— — — — — 9/5/2016 8:57:55 AM — — — — —

Welcome to Minitab, press F1 for help.
```

Test and CI for One Proportion

```
Sample    X     N     Sample p         95% CI
1         10   67     0.149254    (0.073965, 0.257402)
```

The result is that Francona would have been 95% confident that Ortiz's true average in 2010 was going to be between 0.074 and 0.257, horrible numbers. But Ortiz responded to his mangers faith and hit 0.270 with 32 homers and 102 RBIs.

PROPORTION EXERCISE

Deshaun Watson of Clemson is one of the top five collegiate football players in the country. But the junior quarterback has been trending down over his three seasons. For instance:

Stat	2014	2015	2016
NCAA QB Rating	88.6	156.3	149.2
QBR (ESPN)	87.1	77.9	66.3
Yards per attempt	10.7	8.36	7.88
Completion percentage (%)	67.9	67.8	64.4

While his numbers are down, let's test them to see if statistically they really are down. So far we only know how to test a proportion (and just one proportion at that). So to see if his performance is declining, we can only test a proportion. Looking at the four stats above, which is a proportion?

Right! Completion percentage! When you see a fraction, or a decimal, or a percent that is a proportion. So the only stat we can test so far is completion percentage. Last year he completed 67.8% of his passes. This year (2016), he has completed 204 out of 317 passes.

Do a hypothesis test and a confidence interval to determine if Deshaun Watson's performance is declining this year compared to last year.

Hypothesis Test
1. H_o:
2. H_A:
3. Test statistic and its value:
4. p-value:
5. Your statistical conclusion:
6. Your interpretation:

Confidence Interval
7. 95% Confidence interval:
8. Your statistical conclusion:
9. Your interpretation:

Confirmation
10. Does the confidence interval confirm the results of the hypothesis test and why?

ANSWER

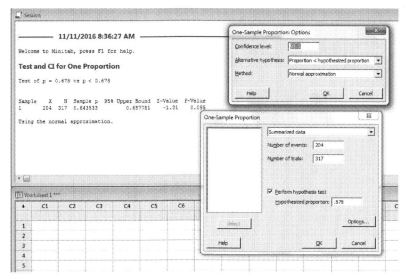

Hypothesis Test

1. **H_0: p = .678** (remember to use last year's number as a decimal, not 67.8%)
2. **H_A: p < .678** (Choose < because the sample P above is .64 which is < .678)
3. **Test statistic and its value:** z = -1.31
4. **p-value:** 0.095
5. **Your statistical conclusion:** Since the *p*-value is > 0.05, I fail to reject the H_0
6. **Your interpretation:** We could not reject 0.678, so he could still be 0.678 in 2016. His performance is not declining.

Confidence Interval

7. **95% Confidence interval:** (0.59081, 0.69626)
8. **Your statistical conclusion:** I am 95% confident Watson will complete between 59% and almost 70% of his passes in 2016.
9. **Your interpretation:** Since 0.678 falls in the interval, he could still be a 0.678 passer.

Confirmation

10. **Does the confidence interval confirm the results of the hypothesis test and why?** Yes, both tests show that he could still be a 67.8% passer, so his performance is not declining.

Hypothesis testing

PREVIEW: WE'LL USE THE HYPOTHESIS TEST TO DOUBLE CHECK OUR NEW TEST

We have done the six steps of hypothesis testing before, so the concept is not new when we do inference. But in inference we will write the null hypothesis in a different form for each problem. When we do one-sample z-test for proportions problems, we test the null hypothesis. For you as a student, the null will always be the old number. So if David Ortiz batted 0.266 with the Twins, and now he is with the Red Sox, 0.266 becomes the old number. In general, for the null we would say his proportion is still equal to the old number. Specifically, H_0: p = 0.266.

Here's the process.

1. **Decide on the null hypothesis, H_0.**

    The H_0 is equal to the old number p = 0.266.

2. **Decide on the alternative hypothesis, H_A.**

    There are three alternatives to being equal to the old number. It can be less than, not equal, or greater than, <, ≠, or >. So for the alternative, keep p and keep 0.266, but change the = sign to one of the alternatives. The alternative will depend on the problem. For instance, if we think David Ortiz will improve, that is >. We would write that as H$_A$: p > 0.266

3. **Go to Minitab and perform the One Proportion test.** Minitab will calculate the z and the *p*-value.

4. **For the answer, list the H$_O$, the H$_A$, the z, and the *p*-value.**

5. **Identify the statistical result as before.**

   a. if your *p*-value is < 0.05, you reject the H$_O$
   b. if your *p*-value is > 0.05, you fail to reject the H$_O$

6. **Then summarize the results of your test in the context of the problem, interpret the results.**

Example: In 2009, David Ortiz hit 0.238. On May 5, 2010, he was 10 for 67. Is this evidence that he is no longer a 0.238 hitter?

1. **H$_O$:** p = 0.238 (we did that because 0.238 is the old number, and p stands for proportion)

2. **H$_A$:** p < 0.238 (we do less than because 10/67 is 0.149, which is less than 0.238)

   Now we can choose stat, basic stat, 1-proportion, and enter the data as shown below:

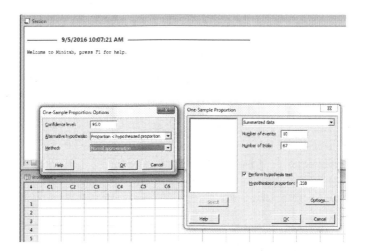

```
————— 9/5/2016 10:07:21 AM —————

Welcome to Minitab, press F1 for help.

Test and CI for One Proportion

Test of p = 0.238 vs p < 0.238

Sample   X    N  Sample p  95% Upper Bound  Z-Value  P-Value
1        10  67  0.149254  0.220860         -1.71    0.044

Using the normal approximation.
```

3. **Test statistic and its value:** z = -1.71
4. **p-value:** 0.044
5. **Your statistical conclusion:** Since the *p*-value < 0.05, I reject the H₀
6. **Your interpretation:** David Ortiz is no longer a 0.238 hitter, he is worse.

Let's try one more example. The Patriots played the Steelers Sunday and beat them with a very efficient passing game. Belichick has always had success in the past against the Steelers, and it turns out his game plan is based on throwing the ball 73% of the time. This Sunday though, the Patriots threw the ball 43 times out of 67 plays, for a proportion of 0.642. This was 64.2% of the plays. Is this evidence that the Patriots and Belichick changed their game plan and they did not rely on the pass as much?

```
————— 9/5/2016 10:18:09 AM —————

Welcome to Minitab, press F1 for help.

Test and CI for One Proportion

Test of p = 0.73 vs p < 0.73

Sample   X    N  Sample p  95% Upper Bound  Z-Value  P-Value
1        43  67  0.641791  0.738142         -1.63    0.052

Using the normal approximation.
```

H_0: p = 0.73 (the old number is 0.73)

H_A: p < 0.73 (use < because the new number of 0.642 is < 0.72)

Test statistic and its value: z = -1.63

p-value: 0.052

Your statistical conclusion: Since the *p*-value > 0.05, I fail to reject the H_0

Your interpretation: Belichick did not change his game plan. We saw differences (0.73 vsS 0.64) but those differences are just due to random variation.

1-proportion questions

PREVIEW: NOW WE'LL PUT BOTH TESTS TOGETHER

Now that we are comfortable with confidence intervals and hypothesis testing, you have no other tests to learn! That is all the tests we will do in sports stats. Those two tests will test the same data, giving us two methods to answer the same question. Most times they will give us the same answer. But sometimes they are different, so as a last step we will check to see if they confirm each other. Here are four examples of 1-proportion questions using both the hypothesis test and the confidence interval to answer them, exactly as you will see on the third exam.

The Boston Bruins power play in 2014 was successful 17.8% of the time. In 2015, the Bruins started the year on fire, with 22 goals in 70 power plays. Did the Bruins power play unit improve in 2015? Let's see.

Go to Minitab. We choose stat, basic stat, 1-proportion. The problem has percents and fractions (22/70) in it, so the problem is describing proportions. Under One Proportion, click summarized data. (The NHL summarized the data for us.) Enter 22 for the number of events, and 70 for the number of trials. If all we wanted was a confidence interval, we would stop here. But, we want the confidence interval and the hypothesis test to answer the question.

So click on hypothesis test. We are going to test the old number to see if the Bruins really have improved. The old number is their power play success rate from last year, or 17.8%. Our problem is that just like earlier problems, Minitab does not like percents because they're a ratio, e.g., 17.8 out of a 100 rather than a single number. To make the percent into a number, we move the decimal two places to the left. So on Minitab you will enter 0.178 next to the hypothesis test.

Now click on options. Notice the confidence level defaults to 95%. That is good. We will use a 95% confidence level all semester. But the next box is your alternative hypothesis. It defaults to ≠. We think the Bruins have improved in 2015, so we want > because we think they are scoring more often on the power play. So click > for the alternative. Finally, Minitab defaults to the exact method, but we want the Normal Distribution, so click normal.

```
Test and CI for One Proportion

Test of p = 0.178 vs p > 0.178

Sample    X    N   Sample p   95% Lower Bound   Z-Value   P-Value
1        22   70   0.314286      0.223019        2.98      0.001
```

You calculated your test statistic and your *p*-value, so let's finish the hypothesis test.

Hypothesis Test
1. **H_0: p = 0.178**
2. **H_A: p > 0.178**
3. **Test statistic and its value: z = 2.98**
4. ***p*-value: 0.001**
5. **Your statistical conclusion:** Since the *p*-value < 0.05, I reject the H_0
6. **Your interpretation:** The Bruins power play is better in 2015

Now we will confirm the hypothesis test with a confidence interval. To get the confidence interval, just go back to options and change > to ≠. The results are below:

```
Test and CI for One Proportion

Test of p = 0.178 vs p ≠ 0.178

Sample  X   N   Sample p         95% CI          Z-Value   P-Value
1       22  70  0.314286(0.205535, 0.423037)     2.98      0.003

Using the normal approximation.
```

Confidence Interval

7. **95% Confidence interval:** (0.21, 0.42)

8. **Your statistical conclusion:** I am 95% confident the Bruins will score on the power play between 21% and 42% of the time this year.

9. **Your interpretation:** Since the entire confidence interval is above 0.178, the Bruins power play is better in 2015.

Confirmation

10. **Do the two tests confirm each other?** Yes, because they both show the Bruins power play is better in 2015.

Here is an example with no change. In 2014, the Washington Capitals had the NHL's best power play at 25.3%. In 2015, they started the season with 19 goals in 74 chances, even better than 2014. Are the Capitals better on the power play in 2015?

Hypothesis Test

1. **H₀:** $p = 0.253$
2. **Hₐ:** $p > 0.253$

```
Test and CI for One Proportion

Test of p = 0.253 vs p > 0.253

Sample  X   N   Sample p   95% Lower Bound   Z-Value   P-Value
1       19  74  0.256757   0.173228          0.07      0.470

Using the normal approximation.

Test and CI for One Proportion

Test of p = 0.253 vs p ≠ 0.253

Sample  X   N   Sample p   95% Lower Bound        Z-Value   P-Value
1       19  74  0.256757   (0.157226, 0.356288)   0.07      0.941

Using the normal approximation.
```

3. **Test statistic and its value:** $z = 0.07$
4. **p-value:** 0.470
5. **Your statistical conclusion:** Since the *p*-value > 0.05, I fail to reject the H₀
6. **Your interpretation:** The Capitals power play has not improved from last year

Confidence Interval

7. **95% Confidence interval:** (0.157, 0.356)
8. **Your statistical conclusion:** I am 95% confident the Capitals will end the year between 15.7% and 35.6% successful on the power play.
9. **Your interpretation:** Since 0.253 is in the interval, they could still be a 25.3% power play unit.

Confirmation

10. **Do the hypothesis test and the confidence interval confirm each other?** Yes, because they both show the Capitals could still be at the level of 2014.

One more example. Patrice Bergeron is usually one of the top face-off men in the NHL. In 2014, he won 60.2% of his face-offs. But in 2015 he slipped, only winning 311 out of 577 face-offs. Is Patrice Bergeron no longer an elite face-off man?

Hypothesis Test

1. H_0: p=0.602
2. H_A: p< 0.602 (the wording is "he slipped" and "no longer elite," so <)

```
Test and CI for One Proportion

Test of p = 0.602 vs p < 0.602

Sample   X    N    Sample p   95% Upper Bound   Z-Value   P-Value
1       311  577   0.538995       0.573129       -3.09     0.001

Test and CI for One Proportion

Test of p = 0.602 vs p ≠ 0.602

Sample   X    N    Sample p       95% CI          Z-Value   P-Value
1       311  577   0.538995  (0.498322,0.579668)   -3.09     0.002

Using the normal approximation
```

3. **Test statistic and its value:** z = -3.09
4. **p-value:** 0.001
5. **Your statistical conclusion:** Since the *p*-value < 0.05, I reject the H_0
6. **Your interpretation:** Bergeron has slipped, he is no longer an elite face-off man

Confidence Interval

7. **95% Confidence interval:** (0.498, 0.580)
8. **Your statistical conclusion:** I am 95% confident Bergeron will finish the year winning between 49.8% and 58% of his face-offs.
9. **Your interpretation:** Yes because they both show he is no longer a 60.2% player, he is something less than that

Confirmation

10. **Do the two tests confirm each other?** Since 0.602 is not in the interval Bergeron is no longer a 60.2% player, he is something less than that.

1-PROPORTION EXAMPLE

The Boston Celtics were 21-20 at home in the 2014/15 season.

Calculate a 95% confidence interval using last year's data to predict their home win/loss percentage this year.

```
Test and CI for One Proportion

Sample X    N    Sample p           95% CI
1      21   41   .512195      (0.351342, 0.671221)
```

Notice for the number of trials (N above) I used 41 because that is the total number of games (trials) for the Celtics at home. That gave me the confidence interval. So I am 95% confident the Celtics will win between 35% and 67% of their home games this year.

In their history, the Celtics have been a ..696 team at home. Could they still be a 0.696 team this year based on last year's team? Use the hypothesis test to answer the question and calculate a confidence interval to confirm your result.

```
Test and CI for One Proportion

Test of p = 0.696 vs p < 0.696

Sample  X   N   Sample p   95% Upper Bound   Z-Value   P-Value
1       21  41  0.512195   0.640598          -2.56     0.005

Using the normal approximation.

Test and CI for One Proportion

Test of p = 0.696 vs p ≠ 0.696

Sample  X   N   Sample p       95% CI         Z-Value   P-Value
1       21  41  0.512195 (0.359193, 0.665197) -2.56     0.011

Using the normal approximation.
```

Hypothesis Test

1. **H_0: p = 0.696**
2. **H_A: p < 0.696**
3. **Test statistic and its value:** z = -2.56
4. **p-value:** 0.005
5. **Your statistical conclusion:** Since the *p*-value < 0.05, I reject the H_0
6. **Your interpretation:** The Celtics are no longer a 0.696 team at home, they are worse.

Confidence Interval

7. **95% Confidence interval:** (0.359, 0.665)
8. **Your statistical conclusion:** I am 95% confident the Celtics will win between 36% and 66.5% of their home games this year.
9. **Your interpretation:** Since 0.696 is not in the interval, the Celtics will not win 69.6% of their home games, they will be worse (because the entire interval is below 0.696)

Confirmation

10. **Do the two tests confirm each other?** Yes, they both show the Celtics will be worse at home this year.

1-PROPORTION EXERCISES

In 2014, Dustin Pedroia led the Red Sox in at bats, yet he had 153 hits (out of those 551 at bats), dropping his career numbers to under 0.300 (0.299 specifically). Create a 95% confidence interval to predict his batting average for 2015 from his 2014 stats, and answer the question, could he still be a 0.299 hitter in 2015?

Hypothesis Test
1. H_0:
2. H_A:
3. Test statistic and its value:
4. *p*-value:
5. Your statistical conclusion:
6. Your interpretation:

Confidence Interval
7. 95% Confidence interval:
8. Your statistical conclusion:
9. Your interpretation:

Confirmation
10. Do your results in the hypothesis test confirm the confidence interval? Why?

So far in the 2014 season, new addition A. J. Pierzynski is leading the Red Sox regulars in batting with 25 hits in 86 at-bats. His career average is 0.283. Perform a hypothesis test and a 95% confidence interval to answer the question: Is his 25/86 start with the Red Sox evidence that he is a better hitter with his new team?

Hypothesis Test
1. H_0:
2. H_A:
3. Test statistic and its value:
4. *p*-value:
5. Your statistical conclusion:
6. Your interpretation:

Confidence Interval
7. 95% Confidence interval:
8. Your statistical conclusion:
9. Your interpretation:

Confirmation
10. Does the confidence interval confirm the hypothesis test? Why?

Chris Paul was rated the number 1 point guard in the NBA in the latest ESPN rankings. We will see what happens when he faces Stephen Curry in the Western Conference finals, but that is another story. We will examine Chris Paul to see if he has raised his game to a new level.

In his 10-year career, Paul has played for two teams. During the first 6 seasons he played with New Orleans, where he shot 0.471 During the last 4 seasons with the Los Angeles Clippers, Paul has made 372/1006 shots. We will perform a hypothesis test and a confidence interval determine if Chris Paul has raised his game with LA or could he still be a 0.471 shooter.

Hypothesis Test
1. H_0:
2. H_A:
3. Test statistic and its value:
4. *p*-value:
5. Your statistical conclusion:
6. Your interpretation:

Confidence Interval
7. 95% Confidence interval:
8. Your statistical conclusion:
9. Your interpretation:

Confirmation
10. Do the results of the confidence interval confirm the hypothesis test?

Significance tests about hypothesis

PREVIEW: NOW WE USE BOTH TESTS TO TEST A 1-SAMPLE MEAN

When it is operating correctly, a machine for manufacturing tennis balls produces balls with a mean weight of 57.6 grams. The last eight balls manufactured had weights of:

57.3   57.4   57.2   57.5   57.4   57.1   57.3   57.0

Is the machine operating correctly?

This question is a little different from those on the previous pages, there is no fraction, no percent, and no decimal even. So it is not a proportion problem. It is a 1-sample mean problem. You can see we have 1 sample of 8 tennis balls. We want to test that sample to see if it meets the requirements of the USTA, which is 57.6g. So easy, right? All we have to do is find the mean of these 8 balls.

Well, we have a problem. We do not know anything about the rest of the tennis balls produced by the machine. We only know about these 8. There might be thousands more. So we do not know the standard deviation of those tennis balls. If we did, we would not need to do the test! But we still have to test, and we do not know the population standard deviation. So we cannot use the z test statistic.

But we have a substitute. It is called the t test statistic. And it looks and acts very much like z, so it is easy to substitute. You just need to remember when you do 1 mean like this problem, or 2 means later on, use t, not z.

A t-test is commonly used to determine whether the mean of a population significantly differs from a specific value (called the hypothesized mean) or from the mean of another population.

Minitab: Put the tennis ball data into C1. Then choose stat, basic stats, 1 sample t. This time the default is OK, we have our data in a column, not summarized. So put the curser in the box and double click on C1, which I called "Tennis Ball Weight". That will move it over. Then yes, we want to do a hypothesis test, so click that and put in our hypothesized value, 57.6. Then click options. The option choices are just like before with proportions, except we do not have the choice of normal because we cannot use z. Click the options and here are the results:

—————— 9/5/2016 12:04:13 PM ——————

Welcome to Minitab, press F1 for help.

One-Sample T: Tennis Ball weight

Test of μ = 57.6 vs < 57.6

Var	N	Mean	StDev	SE Mean	95% Upper Bound	T	P
Weight	8	56.775	1.414	0.500	57.722	-1.65	0.071

One-Sample T: Tennis Ball weight

Test of μ = 57.6 vs ≠ 57.6

Var	N	Mean	StDev	SE Mean	95% CI	T	P
Weight	8	56.775	1.414	0.500	(55.593, 57.957)	-1.65	0.143

Here is how you will answer the same 10 questions:

Hypothesis Test

1. **H_o:** m = 57.6 (notice we use m for mean since we no longer have a proportion)
2. **H_A:** m < 57.6 (use < because the mean weight of the 8 balls was 56.775 grams)
3. **Test statistic and its value:** t = -1.65 (no more z)
4. **p-value:** 0.071
5. **Your statistical conclusion:** Since the *p*-value is > 0.05, I fail to reject the H_o
6. **Your interpretation:** As far as we know, the tennis ball will weigh 57.6g

Confidence Interval

7. **95% Confidence interval:** (55.6g, 58g)
8. **Your statistical conclusion:** I am 95% confident the tennis balls weight between 55.6g and 58 g.
9. **Your interpretation:** Since the CI contains 57.6 grams, the tennis balls could still be 57.6g

Confirmation

10. **Do the two tests confirm each other?** Yes, they both show the tennis balls are good, they could still be 57.6g

ONE SAMPLE T TEST EXAMPLES

I paid $350 for my mountain bike several years ago. Here are current prices.

Brand and Model	Price
Trek VRX 200	1000
Cannondale SuperV400	1100
GT XRC-4000	940
Specialized FSR	1100
Trek 6500	700
Specialized Rockhop	600
Harold Escape A7.1	440
Giant Yukon SE	450
Mongoose SX 6.5	550
Diamondback Sorrento	340
Motiv Rockridge	180
Huffy Anorak 36789	140

Is this evidence that the price of mountain bikes has increased? Why or why not?

Perform a hypothesis test to verify your answer.

Hypothesis Test

1. H_0: m = 350 (the old number)
2. H_A: m > 350 (the question says "has increased," so >)
3. **Test statistic and its value:** t = 2.82
4. ***p*-value:** 0.008
5. **Your statistical conclusion:** Since the *p*-value < 0.05, I reject the H_0
6. **Your interpretation:** Mountain Bike costs have increased

Confidence Interval

7. **95% Confidence interval:** ($411.40, $845.30)
8. **Your statistical conclusion:** I am 95% confident that mountain bikes now cost between $411.40 and $845.30.
9. **Your interpretation:** Since $350 is not in the interval, mountain bike prices have gone up.

Confirmation

10. **Do the two tests confirm each other?** Yes, they both show mountain bike prices have increased.

During the Celtics most recent championship run, Kevin Garnett averaged 8.9 rebounds per game during the regular season. In the first round, the Celts swept the Knicks in four games and Garnett had 12, 10, 12 and 10 rebounds in each game. Did Garnett step up his game in the postseason?

One-Sample T: Kevin Garnett Rebounds

```
Test of μ = 8.9 vs > 8.9

Variable   N  Mean    StDev  SE Mean  95% Lower Bound    T      P
Rebounds   4  11.000  1.155  0.577    9.641              3.64   0.018
```

One-Sample T: Kevin Garnett Rebounds

```
Test of μ = 8.9 vs ≠ 8.9

Variable   N  Mean    StDev  SE Mean    95% CI            T      P
Rebounds   4  11.000  1.155  0.577    (9.163, 12.837)    3.64   0.036
```

Wow, even with just a four game sample, it is enough to show KG stepped up his game in the playoffs. Let's see:

Hypothesis Test

1. **H₀:** m = 8.9 (the old number)
2. **Hₐ:** m > 8.9 ("step up his game")
3. **Test statistic and its value:** t = 3.64
4. **p-value:** 0.018
5. **Your statistical conclusion:** Since the *p*-value < 0.05, I reject the H₀
6. **Your interpretation:** KG stepped up his game in the playoffs

Confidence Interval

7. **95% Confidence interval:** (9.2, 12.8)
8. **Your statistical conclusion:** I am 95% confident that Garnett will get between 9.2 and 12.8 rebounds per game in the playoffs
9. **Your interpretation:** Since 8.9 is not in the interval and the entire interval is higher, KG has stepped up his game in the playoffs.

Confirmation

10. **Do the two tests confirm each other?** Yes, they both show Garnett did better in the playoffs.

ONE SAMPLE T TEST EXERCISE

Now you try it with Stevan Ridley. In Stevan Ridley's rookie season with the Patriots, he rushed for 27.6 yards per game. In year 2, he rushed for: 125, 71, 37, 106, 151, 34, 65, 127, 98, and 28 yards in his first 10 games. Has Ridley's production increased in his second season?

Hypothesis Test
1. H_0:
2. H_A:
3. Test statistic and its value:
4. *p*-value:
5. Your statistical conclusion:
6. Your interpretation:

Confidence Interval
7. 95% Confidence interval:
8. Your statistical conclusion:
9. Your interpretation:

Confirmation
10. Does the confidence interval test confirm the hypothesis test?

2-sample means

PREVIEW: NOW WE USE BOTH TESTS TO TEST A 2-SAMPLE MEAN

In football, the big guys play on the line. But which guys are bigger, the offensive line or the defensive line?

Using the 2010 Patriots for our sample, we can see if there is a difference between the two lines. Notice we are now dealing with two samples, in this case, offensive AND defensive linemen. Here are the weights of the linemen on the New England Patriot's 2010 roster:

Defensive Line	Offensive Line
330	313
260	296
305	310
310	305
310	310
330	300
325	295
295	315
	290

We want to compare the two samples and determine if there is a difference in weight. Our first step is to perform a hypothesis test, then hopefully confirm those results with a confidence interval test. The H_o (null hypothesis) is always that there is no difference. In this case, that the Patriots linemen weigh the same on both sides of the ball. If we do see a difference, that difference is just due to random variation. So our null hypothesis is that the mean weight of the DL will equal the mean weight of the OL.

The alternative hypothesis (H_A) is that there is a difference in weight. The mean weight of the DL does not equal the mean weight of the OL.

The test statistic we will use for two sample means is t. Notice when we get into Minitab when we did one sample we had to choose between z and t for our test statistic. With two samples we have no choice, we have to pick t. So we enter the data into Minitab as shown on the next page and then run a two sample t test.

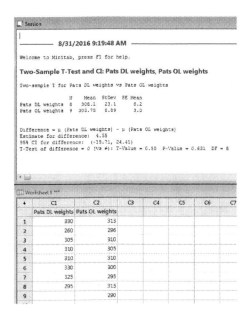

Now we can write up the hypothesis test:

Hypothesis Test
1. **H₀:** Mean weight$_{DL}$ = Mean weight$_{OL}$ (DL first since we loaded them into minitab first)
2. **H$_A$:** Mean weight$_{DL}$ ≠ Mean weight$_{OL}$
3. **Test statistic and its value:** t = 0.50
4. **p-value:** 0.631
5. **Your statistical conclusion:** Since the p-value > 0.05, I fail to reject H₀
6. **Your interpretation:** There is no difference in weights for the DL and the OL.

That's it! You are done with the first test! But as you look at the print out above, it shows the DL mean weight is over 308 pounds, and the OL is only 303.78 pounds. So the weights are different, right? The answer is not statistically. We have decided they are "the same" so the differences we see are due to random variation. In 2011, they OL could weigh more!

Now we confirm our interpretation that there is no difference in weights, the differences we see are due to random variation by performing a confidence interval. The confidence interval is displayed on the Minitab results above, and is (-15.71, 24.41). What does that mean? Since we entered the DL first, we will talk about the result in terms of the DL. The confidence means that the DL weighs between 15.71 pounds less (because of the negative) and 24.41 pounds more (because of the positive) than the OL. Because it could weight less, or it could weigh more, we conclude there is no difference in weight.

Note: If the confidence interval had been negative to negative, or even positive to positive, we would conclude there was a difference in weight (negative to negative would mean the DL weighs less than the OL, and positive to positive would have meant the DL weighs more than the OL). Another way to look at it is anytime you have a negative to a positive, that CI will include zero. And zero means there is a difference of 0 in terms of weight. In other words, no difference. So any CI for two means that contains 0 we can conclude there is no difference, and any CI that does not contain 0 there is a difference.

Now we can add the confidence interval to our hypothesis test and we are done!

Hypothesis Test

1. **H_0:** Mean weight$_{DL}$ = Mean weight$_{OL}$
2. **H_A:** Mean weight$_{DL}$ ≠ Mean weight$_{OL}$
3. **Test statistic and its value:** t = 0.50
4. ***p*-value:** 0.631
5. **Your statistical conclusion:** Since the *p*-value > 0.05, I fail to reject H_0
6. **Your interpretation:** There is no difference in weights for the DL and the OL.

Confidence Interval

7. **95% Confidence interval:** (-15.71, 24.41)
8. **Your statistical conclusion:** I am 95% confident the Pats DL weighs between 15.71 pounds less and 24.41 pounds more than the OL in 2010.
9. **Your interpretation:** Since the interval goes from negative to positive, there is no difference in weight.

Confirmation

10. **Do the two tests confirm each other?** The CI confirms the HT as they both show there is no difference in weights for Pats linemen in 2010.

2 SAMPLE MEAN EXAMPLE

The Springfield College men's volleyball team has won the last 3 National Championships and 9 overall. This year they made it to the championship game again, but fell short in the finals. They can compete with any Division I team in the country, so how do we compare in height? Loyola-Chicago is the number one team in Division I, so we will compare the heights of the SC athletes to Loyola-Chicago's using a hypothesis test and a confidence interval to see if Springfield is shorter than the number one ranked team in the country. The heights for Minitab are on the next page.

```
———————— 9/5/2016 1:55:36 PM ————————
Welcome to Minitab, press F1 for help.

Two-Sample T-Test and CI: Springfield College Mens VB, Loyola Chicago Mens VB

Two-sample T for Springfield College Mens VB vs Loyola Chicago Mens VB

                           N    Mean   StDev  SE Mean
Springfield College Mens  20   74.60    2.93     0.65
Loyola Chicago Mens VB    16   76.81    3.43     0.86

Difference = μ (Springfield College Mens VB) - μ (Loyola Chicago Mens VB)
Estimate for difference:  -2.21
95% CI for difference:  (-4.42, -0.01)
T-Test of difference = 0 (vs ≠): T-Value = -2.05  P-Value = 0.049  DF = 29
```

	C1	C2	C3	C4	C5	C6
	Springfield College Mens VB	Loyola Chicago Mens VB				
4	76	75				
5	74	71				
6	76	77				
7	74	76				
8	73	76				
9	77	78				
10	73	77				
11	75	83				
12	76	78				
13	77	78				
14	69	82				
15	76	75				
16	67	82				
17	79					
18	74					

Hypothesis Test

1. H_0:
2. H_A:
3. Test statistic and its value:
4. *p*-value:
5. Your statistical conclusion:
6. Your interpretation:

Confidence Interval

7. 95% Confidence interval:
8. Your statistical conclusion:
9. Your interpretation:

Confirmation

10. Do the two tests confirm each other?

Volleyball Team

Springfield College		Loyola Chicago
79	72	73
74	76	76
74	76	72
74	76	75
73		71
77		77
73		76
75		76
76		78
77		77
69		83
76		78
67		78
79		82
74		75
75		82

ANSWER

Hypothesis Test
1. H_0: SC = Loyola
2. H_A: SC < Loyola
3. **Test statistic and its value:** t = -2.05
4. **p-value:** 0.049
5. **Your statistical conclusion:** Since the *p*-value < 0.05, I reject the H_0
6. **Your interpretation:** There is a difference in heights, SC is shorter

Confidence Interval
7. **95% Confidence interval:** (-4.42, -0.01)
8. **Your statistical conclusion:** I am 95% confident that SC is between 4.4 and 0.01 inches shorter than Loyola.
9. **Your interpretation:** Because the interval goes from negative to negative, there is a difference, and the difference is SC is shorter.

Confirmation
10. **Do the two tests confirm each other?** Yes, because they both show there is a difference in heights, that SC is shorter.

2 SAMPLE MEAN EXERCISES

In baseball, the American League allows a Designated Hitter (DH) to bat for the pitcher, who is typically a weak hitter. In the National League, the pitcher must bat. The common belief is that this results in American League teams scoring more runs. In inter-league play, when American League teams visit National League teams, the pitcher must bat. So, if the DH does result in more runs, we would expect American League teams to score fewer runs when visiting National League parks.

To test this claim, a random sample of runs scored by American League teams with and without their DH is given in the following table: (ESPN.com)

National League Park Runs	American League Park Runs
1 5 5 4 7	6 2 3 6 8
2 6 2 9 2	1 3 7 6 4
8 8 2 10 4	4 12 5 6 13
4 3 4 1 9	6 9 5 6 7
3 5 1 3 3	4 3 2 5 5
3 5 2 7 2	6 14 14 7 0

You will test the claim that the DH results in more runs scored.

Do a hypothesis test to test the claim that the DH results in more runs scored. Which test statistic will you use, t or z?

Hypothesis Test
1. H_0:
2. H_A:
3. Test statistic and its value:
4. *p*-value:
5. Your statistical conclusion:
6. Your interpretation:

Confidence Interval
7. 95% Confidence interval:
8. Your statistical conclusion:
9. Your interpretation:

Confirmation
10. Do the two tests confirm each other?

The Boston Marathon is one of the top five marathon's in the world. Year in and year out it has provided the best marathoners in the world a chance to compete against each other. As a result, there has been some good rivalries that have developed over time. Today, it is the Kenyans vs. the Ethiopians. Both countries circle Patriots Day on their calendars and send their best runners to Boston.

Google "Boston Marathon results 2016". The BAA site should be near the top. Click on "2016 Top Finishers". Let's test to see which population is better, all the runners from Kenya or all the runners from Ethiopia.

Enter the top four Kenyan times in C1 and the top four Ethiopian times in C2. We will perform hypothesis test and a confidence interval to see if there is a difference between the two countries. Which test statistic will you use, t or z?

Hypothesis Test
1. H_0:
2. H_A:
3. Test statistic and its value:
4. *p*-value:
5. Your statistical conclusion:
6. Your interpretation:

Confidence Interval
7. 95% Confidence interval:
8. Your statistical conclusion:
9. Your interpretation:

Confirmation
10. Do the two tests confirm each other?

Here are the winning times for the men and the women in 15 New York City marathons:

Year	Men	Women
1992	129.5	144.7
1993	130.1	146.4
1994	131.4	147.6
1995	131.0	148.1
1996	129.9	148.3
1997	128.2	148.7
1998	128.8	145.3
1999	129.2	145.1
2000	130.2	145.8
2001	127.7	144.4
2002	128.1	145.9
2003	130.5	142.5
2004	129.5	143.2
2005	129.5	144.7
2006	130.0	145.1

We will use these 15 races as our sample, with our population being the finishing times for men and women in all marathons. Perform a hypothesis test and a confidence interval to see if there is a difference in marathon times for men and women. Which test statistic will you use, t or z?

Hypothesis Test
1. H_0:
2. H_A:
3. Test statistic and its value:
4. *p*-value:
5. Your statistical conclusion:
6. Your interpretation:

Confidence Interval
7. 95% Confidence interval:
8. Your statistical conclusion:
9. Your interpretation:

Confirmation
10. Do the two tests confirm each other?

Estimates of a population

PREVIEW: NOW WE USE BOTH TESTS TO TEST A 1-SAMPLE MEAN

Below are the new test statistics we encountered in Part 3.

	1 sample		2 sample		3 or more
	mean	proportion	mean	proportion	mean
Confidence interval	t	z	t	z	F
Hypothesis test	t	z	t	z	F

Note: Because we are making estimates of a population, we do not know the specifics of the population. (If we did we wouldn't need to make the estimates!).So when dealing with the mean, we do not know the population standard deviation. That means we cannot use the z test for means. We have to substitute the t test (a very good test in itself, just one level below z). So use t whenever you are dealing with means, and the z test statistic when you are dealing with proportions.

PROPORTION EXERCISES

Stephen Curry set an NBA record with 13 3 pointers in a 2016 game against the New Orleans Pelicans.

He set this record after going 0 for 10 from the 3-point range just the game before.

We would think his 0-for-10 would hurt his 3-point percentage, so let's test his 3-point percentage to see if he has gone down this season. In 2015/16, Curry shot an incredible .454 from 3-point range. This year he has made 32 of 71 attempts. Do a hypothesis test and a confidence interval to see if his percentage has suffered.

Hypothesis Test
1. H_O:
2. H_A:
3. Test statistic and its value:
4. *p*-value:
5. Your statistical conclusion:
6. Your interpretation:

Confidence Interval
7. 95% Confidence interval:
8. Your statistical conclusion:
9. Your interpretation:

Confirmation
10. Does the confidence interval confirm the results of the hypothesis test and why?

Tuukka Rask makes 32 saves to lead the Boston Bruins to a 4-0 shutout of the Buffalo Sabres. This game solidifies Rask's spot as a top NHL goaltender this season. He is in the top 5 for goals against average, save percentage, and wins.

Last year he was not an elite goalie. For instance, he finished in a tie for 27th in save percentage last year with a .915 save percentage. This year he has saved 225 out of 239 shots. Let's test to see if he has improved this year. Do a hypothesis test and a confidence interval to see if his save percentage has improved.

Hypothesis Test
1. H_0:
2. H_A:
3. Test statistic and its value:
4. *p*-value:
5. Your statistical conclusion:
6. Your interpretation:

Confidence Interval
7. 95% Confidence interval:
8. Your statistical conclusion:
9. Your interpretation:

Confirmation
10. Does the confidence interval confirm the results of the hypothesis test and why?

2-proportion questions

PREVIEW: NOW WE'LL SEE HOW BOTH TESTS TOGETHER WORK WITH 2 PROPORTIONS

During the 2004 season, San Francisco Giant Barry Bonds was the poster child for steroid abuse. It made him such a dangerous hitter that many teams simply chose to walk him rather than throw him a pitch he could hit. Just before a series in New York, an analyst advised the Mets that they should pitch to Bonds.

As evidence, the analyst reported that thus far in the season, the Giants had scored in 37 of the 79 innings when Bonds was walked intentionally. On the other hand, when opponents did not walk him intentionally, the Giants scored in 107 out of 298 innings.

Does this provide evidence that the Mets should pitch to Bonds as the analyst advises?

We will perform a hypothesis test and a confidence interval to answer the question, and I will load the answers right into Minitab below.

A note about the confidence intervals. They are interpreted exactly as we interpreted the two sample mean intervals in the last few chapters. But, because they are proportions, we read them a little differently. As a proportion, we move the decimal point two places to the right and talk about them as percentages.

To enter two proportion problems into Minitab, go to "stat". Scroll over to "basic stat" and scroll down to two proportion. Click two proportion. In the drop down menu, choose summarized data (the analyst summarized it into the two proportions). Since he claims pitching to Bonds is better, we will enter that first. After entering the two samples, just hit OK. And here is your result:

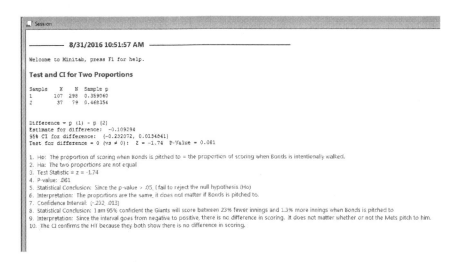

So, in our results above, the estimate for the difference was -0.109294. We understand that as saying the Giants scored almost 11% fewer times when Bonds was pitched to. This is all the analyst saw when he recommended teams pitch to Bonds. He did not look at the CI which was (-0.232072, 0.0134841). We already know that when an interval goes from negative to positive (and therefore contains 0), there is no difference in Giants scoring when Bonds is pitched to or walked intentionally. But the interpretation we talk about in terms of percent's (remember, p and p, proportions and percent's). We interpret the CI as the Giants score between 23% fewer times and 1.3% more times when Bonds is pitched to.

2-PROPORTION EXAMPLES

In 2015, Rick Porcello pitched his first season with the Red Sox and he responded with a 9-15 record. Then in 2016 he wins the Cy Young award! In 2016 he was 22-4. Perform a hypothesis test and a confidence interval to answer the question: Was he significantly better in 2016 or was this season just random variation at work?

Hypothesis Test
1. H_0:
2. H_A:
3. Test statistic and its value:
4. *p*-value:
5. Your statistical conclusion:
6. Your interpretation:

Confidence Interval
7. 95% Confidence interval:
8. Your statistical conclusion:
9. Your interpretation:

Confirmation
10. Does the confidence interval confirm the hypothesis test?

Springfield, Mass. - November 15, 2016 - The Springfield College men's basketball team opened up the 2016 campaign with an 80-69 win at crosstown rival Western New England on Tuesday evening. At the same time, the women beat WNE in double OT, 83-75. So both are 1-0 to start the season, but which team is the better shooting team? The men shot lights out, going 28 for 55 from the floor. The women, not so much. They only hit 23 out of 69 in their win. So perform a hypothesis test and a confidence interval to see if the men were significantly better than the women.

Hypothesis Test
1. H_0:
2. H_A:
3. Test statistic and its value:
4. p-value:
5. Your statistical conclusion:
6. Your interpretation:

Confidence Interval
7. 95% Confidence interval:
8. Your statistical conclusion:
9. Your interpretation:

Confirmation
10. Does the confidence interval confirm the hypothesis test?

ANSWERS

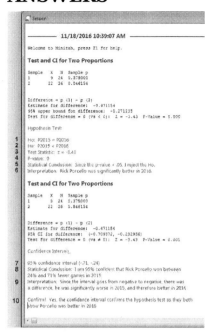

2-PROPORTION EXERCISES

A study by the National Athletic Trainers Association surveyed 1679 high school freshmen and 1366 high school seniors in Illinois. Results showed that 34 of the freshmen and 24 of the seniors had used anabolic steroids.

Is there a significant difference between the proportions of freshmen and seniors who have used steroids? We will perform a hypothesis test and a confidence interval to answer the question.

Hypothesis Test
1. H_o:
2. H_A:
3. Test statistic and its value:
4. p-value:
5. Your statistical conclusion:
6. Your interpretation:

Confidence Interval
7. 95% Confidence interval:
8. Your statistical conclusion:
9. Your interpretation:

Confirmation
10. Do the two tests confirm each other?

When games were sampled from throughout a season, it was found that the home team won 127 of 198 NBA games, and the home team won 57 of 99 NFL games.

Does there appear to be a significant difference between the proportions of home wins? We will perform a hypothesis test and a confidence interval to answer that question.

Hypothesis Test
1. H_0:
2. H_A:
3. Test statistic and its value:
4. p-value:
5. Your statistical conclusion:
6. Your interpretation:

Confidence Interval
7. 95% Confidence interval:
8. Your statistical conclusion:
9. Your interpretation:

Confirmation
10. Do the two tests confirm each other?

A study of injuries to in-line skaters used data from the National Electronic Injury Surveillance System, which collects data from a random sample of hospital emergency rooms. The researchers interviewed 161 people who came to emergency rooms with injuries from in-line skating. Wrist injuries were the most common, so the interviewers broke down the data among those patients who wore wrist guards and those who did not wear wrist guards.

Of the 53 people wearing wrist guards, 6 had wrist injuries. Of the 108 who did not wear wrist guards, 45 had wrist injuries. Is this evidence that skaters who wear wrist guards suffer fewer injuries?

Hypothesis Test
1. H_0:
2. H_A:
3. Test statistic and its value:
4. p-value:
5. Your statistical conclusion:
6. Your interpretation:

Confidence Interval
7. 95% Confidence interval:
8. Your statistical conclusion:
9. Your interpretation:

Confirmation
10. Do the two tests confirm each other?

In class, we discovered that white female student-athletes (498/796) had a higher on-time graduation rate than their male (878/1625) counterparts. Does that also pertain to black female student-athletes?

The NCAA conducted a study and found 54 out of 143 black female student-athletes graduated on-time, while 197 out of 660 black male student-athletes graduated in a timely manner. Perform a hypothesis test and a 95% confidence interval to determine if there is a difference in their on-time graduation rates.

Hypothesis Test
1. H_0:
2. H_A:
3. Test statistic and its value:
4. *p*-value:
5. Your statistical conclusion:
6. Your interpretation:

Confidence Interval
7. 95% Confidence interval:
8. Your statistical conclusion:
9. Your interpretation:

Confirmation
10. Do the two tests confirm each other?

2 PROPORTION AND 2 SAMPLE MEANS EXERCISES

The Springfield College basketball season is underway, so we can look at their point totals to see which is the higher scoring team, the women or the men?

SC Women points scored 2016	SC Men points scored 2016
83	80
46	83
47	74
85	84
85	63
	64

The Boston Celtics last year made 717/2142 3 point shots. This year they have made 222/616 3 point shots so far. Are they a better 3 point shooting team this year?

Which team is taller at Springfield College, the 2016 men's basketball or volleyball team? Heights are in inches.

Basketball		Volleyball	
76	76	79	77
74	78	76	78
71	80	78	75
76		74	73
72		72	79
70		77	74
74		73	78
79		66	72
72		76	76

ANOVA

PREVIEW: HERE IS A TEST THAT HELPS WITH THREE OR MORE MEANS

We began the section on inference by investigating samples with just one mean. Then we looked at comparing two means. But what if we want to compare 3 means? 4 means? 10 means? To investigate one mean we used the t test statistic because we did not know the population standard deviation. And the same with 2 means, we used the t test. Well, one way to compare more than two means is to do multiple t tests. But multiple t tests give us too much error, specifically, type I error. So we want to do one test to compare however many means we have.

That one test is called an Analysis of Variance (ANOVA). As you can guess, we analyze the variances associated with each mean. But now we need a new test statistic to replace t. ANOVA uses the F test statistic to see if there are any differences between the means we are studying. The formula to determine F and determine if the variation between the sample means is significant is:

$$F = \frac{\text{Variation Among Sample Means}}{\text{Variation Among Individuals in Each Sample}}$$

There are many uses for ANOVA but we will keep it simple. We will use only One-way between groups ANOVA. This will let us test the difference between groups. A typical use would be to determine if there is a difference between salaries in the four major professional sports (see ANOVA Example). The groups will be the NFL, NHL, MLB and NBA. We will enter the salaries into Minitab, then run the one-way ANOVA to see if there are any differences.

We will use the same tests as we have all along, hypothesis tests and confidence intervals.

For the hypothesis test,

H₀: Mean$_{NFL}$ = Mean$_{NHL}$ = Mean$_{MLB}$ = Mean$_{NBA}$
H$_A$: At least one of the means is different.
Test Statistic: F.
Then the same ***p*-value** as before.
With the same statistical conclusion and interpretation.

For the confidence intervals, we will have four in our example.
But then the same statistical conclusion and interpretation.

Then we check to see if the two tests confirm each other.

Note: The F test will only tell us if there is a difference, not where the difference is. Usually we would do post-hoc comparisons to see where the differences lie. But those techniques are upper level stats. So we will, with caution, use the confidence intervals. If the confidence intervals overlap, there is no difference. If at least one does not overlap, we will conclude there is a difference, and we will know which mean is different.

ANOVA EXAMPLE

I copied the top 100 paid athletes for each of the four major sports into C1 through C4 in Minitab. Then I clicked Stats, and notice ANOVA has its own link. I click ANOVA, then nice basic one-way. ANOVA defaults to having the data in one column, I entered it in four different columns, so I change that. Now in the responses box, I put all four leagues. Then I click OK. And here is the result:

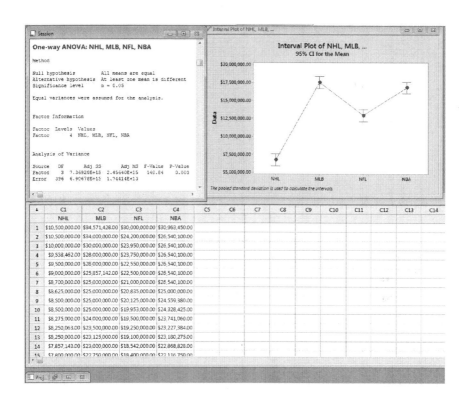

So let's do the Hypothesis Test first to determine if there is a difference in salaries for the four major leagues. So we are comparing mean salaries in each league.

Hypothesis Test

1. **H₀:** Mean$_{NHL}$ = Mean$_{MLB}$ = Mean$_{NFL}$ = Mean$_{NBA}$
2. **H$_A$:** At least one of the means is different.
3. **Test statistic and its value:** F: 140.84
4. **p-value:** 0
5. **Your statistical conclusion:** Since the *p*-value < 0.05, I reject the H₀
6. **Your interpretation:** At least one of the leagues is different

Now we will do the confidence intervals:

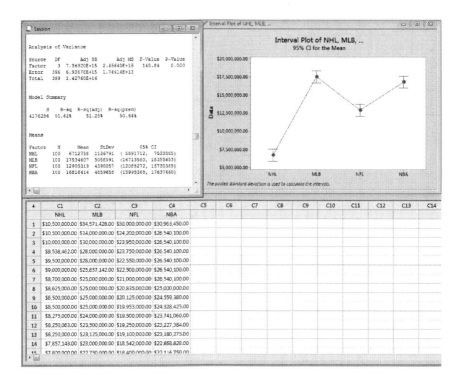

Everything is the same but I scrolled down in the session window to show the actual confidence intervals. Now we can complete our analysis and interpretation.

Confidence Interval

7. **95% Confidence interval:**
   I am 95% confident NHL players make between $5.9M and $7.5M
   I am 95% confident MLB players make between $16.7M and $18.4M
   I am 95% confident NFL players make between $12.1M and $13.7M
   I am 95% confident NBA players make between $16.0M and 17.6M

8. **Your statistical conclusion:** MLB and the NBA overlap so there is no difference between them, but the NHL and the NFL do not overlap so there is a difference between them.

9. **Your interpretation:** Baseball players and basketball players make the most money, but there is no statistical difference between them. They do make more than their counterparts in the NFL, and all three of them make more than the NHL players.

Confirmation

10. **Do the two tests confirm each other?** Yes, they both show that at least one of the means is different.

SPRINGFIELD COLLEGE
Sports Statistics Exam 3 Spring 2015

In 2014, Brock Holt's first full year in the majors, he batted 0.281. Not bad. So far in 2015 though, he is 17 for 44 and leads the Red Sox in several categories. Holt's 17 for 44 is an improvement over 2014, so can we say he is a better hitter this year?

We will perform a hypothesis test and a confidence interval to determine if Holt has improved in 2015 or could he still be a 0.281 hitter at the end of this year.

Hypothesis Test
1. H_0:
2. H_A:
3. Test statistic and its value:
4. p-value:
5. Your statistical conclusion:
6. Your interpretation:

Confidence Interval
7. 95% Confidence interval:
8. Your statistical conclusion:
9. Your interpretation:

Confirmation
10. Do the two tests confirm each other?

The Boston Bruins missed the playoffs in 2015 after being picked by many to win the East. Part of the issue was the goalies, as Tuukka Rask had a down year and Niklas Svedberg played like the back-up he is. So when the dust settled and the season ended, was there any difference between the two goalies?

Tuukka saved 1855 out of 2011 shots, and Niklas Svedberg saved 390 out of 425 shots in 2014/2015. We would expect Tuukka to have a higher save percentage as the number one goalie, but based on the stats was Tuukka better than Svedberg?

Hypothesis Test
1. H_0:
2. H_A:
3. Test statistic and its value:
4. *p*-value:
5. Your statistical conclusion:
6. Your interpretation:

Confidence Interval
7. 95% Confidence interval:
8. Your statistical conclusion:
9. Your interpretation:

Confirmation
10. Do the two tests confirm each other?

The Boston Celtics were swept in the first round of the 2015 NBA playoffs by the Cleveland Cavaliers. The Celtics are a young team so the bright side of the sweep was the experience gained by the younger players. But were the Celtics younger than the Cavaliers? Below are the ages of the 2015 playoff rosters.

We will now perform a hypothesis test and a confidence interval to see if the Celtics are younger team than the Cavaliers.

Celtics	Cavaliers
25	24
29	23
24	35
24	23
27	30
28	34
24	26
24	36
21	35
23	28
26	30
26	24
32	32
19	29
25	24

Hypothesis Test
1. H_o:
2. H_A:
3. Test statistic and its value:
4. p-value:
5. Your statistical conclusion:
6. Your interpretation:

Confidence Interval
7. 95% Confidence interval:
8. Your statistical conclusion:
9. Your interpretation:

Confirmation
10. Do the two tests confirm each other?

Compared to 2013, Brady's passing numbers were down this year, but he still led the New England Patriots to the Super Bowl win following the 2014 season. In 2013, he passed for 271 yards per game. Here are Brady's numbers for the 2014 Super Bowl winning regular season:

We will now perform a hypothesis test and a confidence interval to see if Tom Brady could still be a 271 yards per game quarterback next year, even though his numbers were down this year.

Yards
249
149
234
159
292
361
261
354
333
257
349
245
317
287
182
80

Hypothesis Test

1. H_0:
2. H_A:
3. Test statistic and its value:
4. *p*-value:
5. Your statistical conclusion:
6. Your interpretation:

Confidence Interval

7. 95% Confidence interval:
8. Your statistical conclusion:
9. Your interpretation:

Confirmation

10. Do the two tests confirm each other?

STCC is a local community college, SC is an NCAA Division III college and AIC is an NCAA Division II school, and all are located right here in Springfield. All three have Men's basketball teams competing at three different levels. So are the players any different? We will look at height for our example and ask the question: Is there a difference in heights among the three basketball teams and if so, which is taller?

STCC	SC	AIC
74	73	79
73	73	76
71	74	72
72	75	71
72	75	73
74	70	72
75	75	76
75	79	76
69	76	75
72	76	69
75	76	75
73	79	70
74		79
77		72

Hypothesis Test

1. H_0:
2. H_A:
3. Test statistic and its value:
4. *p*-value:
5. Your statistical conclusion:
6. Your interpretation:

Confidence Interval

7. 95% Confidence interval:
8. Your statistical conclusion:
9. Your interpretation:

Confirmation

10. Do the two tests confirm each other?

SPRINGFIELD COLLEGE
Sports Statistics Exam 3 Fall 2015

The Bruins are the hottest team in the East, but lost Monday night to Edmonton, to fall to 16-9-4 overall. The season has been an anomaly so far as the Bruins have been much better on the road (the Edmonton loss was another home loss). At home they have won only 6/15 games, and on the road they have won 10/14 games. Perform a hypothesis test and a confidence interval to evaluate whether the Bruins winning percentage is really lower at home.

Hypothesis Test
1. H_0:
2. H_A:
3. Test statistic and its value:
4. *p*-value:
5. Your statistical conclusion:
6. Your interpretation:

Confidence Interval
7. 95% Confidence interval:
8. Your statistical conclusion:
9. Your interpretation:

Confirmation
10. Do the two tests confirm each other?

Rob Gronkowski is having the best season of his career in 2015, both in terms of yards per game and yards per catch. We will look at his yards per game and test the data to see if this is his best season ever. Given his yards per game, perform a hypothesis test and a confidence interval to determine if yards per game has improved from last year (74.9).

Yards/Game
94
113
101
67
50
108
113
47
113
37
87

Hypothesis Test
1. H_0:
2. H_A:
3. Test statistic and its value:
4. *p*-value:
5. Your statistical conclusion:
6. Your interpretation:

Confidence Interval
7. 95% Confidence interval:
8. Your statistical conclusion:
9. Your interpretation:

Confirmation
10. Do the two tests confirm each other?

In October 2015, the World Ironman Championships were held in Kona, Hawaii. If you have visited Hawaii, you know how oppressive the heat, but 2015 was even hotter than usual. If it really was hotter, it would show in the marathon times, the last event of the day. Below are the marathon times in minutes for the top ten finishers. Perform a hypothesis test and a confidence interval to see if there was a difference in times this year.

2013	2014	2015
171	174	172
177	171	170
179	168	176
173	168	174
171	172	176
168	173	173
178	171	184
164	176	186
176	170	186
186	180	179

Hypothesis Test
1. H_0:
2. H_A:
3. Test statistic and its value:
4. *p*-value:
5. Your statistical conclusion:
6. Your interpretation:

Confidence Interval
7. 95% Confidence interval:
8. Your statistical conclusion:
9. Your interpretation:

Confirmation
10. Do the two tests confirm each other?

Henrik Lundqvist is one of the top goalies in the NHL, but in 2015 he may be playing better than ever. Last season his save percentage was 0.922, but this year he has saved 715 of 766 shots. Perform a hypothesis test and a confidence interval to answer the question, has Lundqvist improved in 2015?

Hypothesis Test
1. H_0:
2. H_A:
3. Test statistic and its value:
4. *p*-value:
5. Your statistical conclusion:
6. Your interpretation:

Confidence Interval
7. 95% Confidence interval:
8. Your statistical conclusion:
9. Your interpretation:

Confirmation
10. Do the two tests confirm each other?

The Red Sox finished last in 2015, while the Royals won the World Series. Let's compare them to see if there is really any difference between the two teams in terms of defense. The Red Sox finished 12 in the AL in total defense, and the Royals sixth. The Red Sox fielding percentage encompassed 5970 plays/6067 chances, while the Royals made 5967/6055. Perform a hypothesis test and a confidence interval to determine if the Red Sox defense was worse than the Royals.

Hypothesis Test
1. H_0:
2. H_A:
3. Test statistic and its value:
4. p-value:
5. Your statistical conclusion:
6. Your interpretation:

Confidence Interval
7. 95% Confidence interval:
8. Your statistical conclusion:
9. Your interpretation:

Confirmation
10. Do the two tests confirm each other?

SPRINGFIELD COLLEGE
Sports Statistics Exam 3 Spring 2016

This weekend the best rivalry in baseball resumes when the Yankees travel to Fenway to take on the Boston Red Sox. So far this season, the Red Sox have won 12 out of 21 games, and the Yankees have won only 8 out of 20 games. Perform a hypothesis test and a confidence interval to see if there is a statistical difference in their records heading into this first showdown of 2016.

Hypothesis Test
1. H_0:
2. H_A:
3. Test statistic and its value:
4. p-value:
5. Your statistical conclusion:
6. Your interpretation: Is there a difference in the records of the Red Sox and Yankees this year?

Confidence Interval
7. 95% Confidence interval:
8. Your statistical conclusion:
9. Your interpretation: Is there a difference in the records of the Red Sox and Yankees this year?

Confirmation
10. Do the results of the confidence interval confirm your results in your hypothesis test? Why?

Springfield College has been ranked in the top 25 colleges as a place where a female athlete wants to attend and play sports. We will perform a hypothesis test and a confidence interval using ANOVA to compare three women's teams (Volleyball, Basketball and Field Hockey) to see if there is a statistical difference in heights for the athletes.

Hypothesis Test
1. H_0:
2. H_A:
3. Test statistic and its value:
4. *p*-value:
5. Your statistical conclusion:
6. Your interpretation: Is there a difference in heights between the three teams?

Confidence Interval
7. 95% Confidence interval:
8. Your statistical conclusion:
9. Your interpretation: Is there a difference in heights between the three teams?

Confirmation
10. Do the results of the confidence interval confirm your results in your hypothesis test? Why?

2015 Springfield College Women

Volleyball	Basketball	Field Hockey
64	63	66
68	67	64
71	67	60
66	67	64
70	67	61
69	66	63
69	70	63
70	66	65
69	71	67
67	71	63
70	69	67
69	70	63
69	67	62
68	74	64
74		65
72		64
71		64
71		63
		63
		65
		61
		64

So far in the 2016 NBA playoffs, through four games, the leading scorer for all teams is Lebron James? No. Stephen Curry? No. Isaiah Thomas? YES! But is he statistically better than James? Perform a hypothesis test and a confidence interval to determine if Thomas is having a better post season than James.

Hypothesis Test
1. H_0:
2. H_A:
3. Test statistic and its value:
4. *p*-value:
5. Your statistical conclusion:
6. Your interpretation: Is Thomas having a better post season than James?

Confidence Interval
7. 95% Confidence interval:
8. Your statistical conclusion:
9. Your interpretation: Is Thomas having a better post season than James?

Confirmation
10. Do the results of the confidence interval confirm your results in your hypothesis test?

Isaiah Thomas	Lebron James
27	22
16	27
42	20
28	22

Springfield College has the Breanna Stewart of Women's Volleyball, Lauren Holt, one of the best volleyball players in the country, and just a junior. And she keeps getting better. As a sophomore, Lauren scored 10.84 points/match, and led Springfield College to the NCAA's. This year, as a junior, she increased her scoring (see worksheet) and overall game and again led SC into the NCAA's. Use the data in the worksheet (her points scored per match in 2015) to see if she is statistically better than last year.

Lauren Holt
2015-16
Pts/Game

Sept	Oct	Nov
9.5	19.5	15.5
7.5	12.5	10.5
12.5	11.5	13.0
12.5	16.0	10.5
6.0	9.5	10.5
11.5	14.5	
13.0	6.0	
15.0	9.0	
17.0	19.0	
14.5	17.0	
9.5	11.5	
12.5	11.5	
13.0	14.5	
13.5	16.0	
10.0	17.0	

Hypothesis Test

1. H_o:
2. H_A:
3. Test statistic and its value:
4. p-value:
5. Your statistical conclusion:
6. Your interpretation:

Confidence Interval

7. 95% Confidence interval:
8. Your statistical conclusion:
9. Your interpretation:

Confirmation

10. Do the two tests confirm each other?

Tom Brady was suspended for four games for his involvement in Deflategate. Tom Brady had his suspension removed upon appeal. And finally Tom Brady had his four game suspension, and he will be gone for the first four games of the 2016 season. So did his performance suffer in 2015, when he had to use properly inflated balls? In 2014 (with deflated balls), he completed 64.1% of his passes. In 2015 (with properly inflated balls), he completed 402 out of 624 passes. Perform a hypothesis test and a confidence interval to determine if his percentage went down because he had to use footballs that were over inflated compared to last year?

Hypothesis Test
1. H_0:
2. H_A:
3. Test statistic and its value:
4. *p*-value:
5. Your statistical conclusion:
6. Your interpretation:

Confidence Interval
7. 95% Confidence interval:
8. Your statistical conclusion:
9. Your interpretation:

Confirmation
10. Do the two tests confirm each other?

SPRINGFIELD COLLEGE
Sports Statistics Exam 3 Fall 2016

In 1998, Mark McGwire hit 70 home runs to set the Major League Baseball record. That same year Sammy Sosa battled him into September for the HR lead, but came up four short with 66 HR's, still second best of all time. But 3 years later Barry Bonds pushed them both down a rung in the record books with his 73 HR's. After that MLB and the Union agreed to test for PED's, so those records may never be touched.

My question is: Is there a strength difference between those three players and the HR champions in 2016 because of that testing? I looked at home run distance instead of quantity. I entered the home run distances into Minitab. The results on Minitab were:

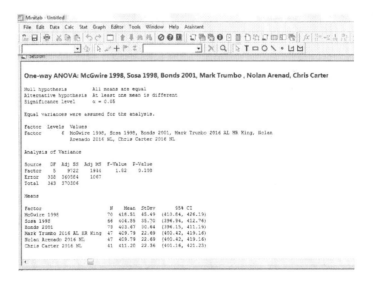

	C1	C2	C3	C4	C5	C6
	McGwire 1998	Sosa 1998	Bonds 2001	Mark Trumbo 2016 AL HR King	Nolan Arenado 2016 NL	Chris Carter 2016 NL
1	360	371	420	390	390	429
2	380	420	370	389	389	434
3	425	350	415	411	411	376
4	450	390	410	404	404	371
5	510	400	320	429	429	408
6	369	388	410	448	448	361
7	430	480	488	389	389	436
8	370	350	442	422	422	374
9	360	460	417	385	385	393
10	370	420	420	406	406	422
11	350	400	436	423	423	401
12	430	380	450	401	401	405

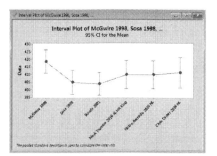

We will perform a hypothesis test and a confidence interval to determine if there is a difference in HR distance between the steroid era and 2016.

Hypothesis Test
1. H_o:
2. H_A:
3. Test statistic and its value:
4. *p*-value:
5. Your statistical conclusion:
6. Your interpretation:

Confidence Interval
7. 95% Confidence interval:
8. Your statistical conclusion:
9. Your interpretation:

Confirmation
10. Do the results of the confidence interval confirm the hypothesis test?

In 2015, the University of Washington football team was 7-6, barely making a bowl game. This year they are 12-1 and headed for the college football playoffs! They will get killed in their first game (Alabama), but at least they made it.

We will perform a hypothesis test and a confidence interval to determine if Washington is statistically better this year.

Hypothesis Test
1. H_o:
2. H_A:
3. Test statistic and its value:
4. *p*-value:
5. Your statistical conclusion:
6. Your interpretation:

Confidence Interval
7. 95% Confidence interval:
8. Your statistical conclusion:
9. Your interpretation:

Confirmation
10. Do the results of the confidence interval confirm the hypothesis test?

3

Since the Celtics obtained Isaiah Thomas, they have a .590 winning percentage with him in the starting line-up. Some radio hosts have claimed the Celtics are better without Thomas in the starting line-up because they are 22-12 in those games.

We will perform a hypothesis test and a confidence interval to determine if the Celtics are better without Thomas in the starting line-up.

Hypothesis Test
1. H_0:
2. H_A:
3. Test statistic and its value:
4. *p*-value:
5. Your statistical conclusion:
6. Your interpretation:

Confidence Interval
7. 95% Confidence interval:
8. Your statistical conclusion:
9. Your interpretation:

Confirmation
10. Do the results of the confidence interval confirm the hypothesis test?

Does the American League, because of the DH, score more runs per game (R/G)? Google American league runs per game, click on the second link, baseball reference.com. Click on 2016, then AL R/G and NL R/G and copy the numbers into Minitab.

We will perform a hypothesis test and a confidence interval to determine if the AL is higher scoring than the NL.

Hypothesis Test
1. H_0:
2. H_A:
3. Test statistic and its value:

4. *p*-value:
5. Your statistical conclusion:
6. Your interpretation:

Confidence Interval
7. 95% Confidence interval:
8. Your statistical conclusion:
9. Your interpretation:

Confirmation
10. Do the results of the confidence interval confirm the hypothesis test?

Russell Westbrook torched the Celtics Sunday with 37 points, leading OKC to the victory. But, he seems to be torching everyone. Last year he averaged 23.5 points per game. Is he even better in 2016? Google Russell Westbrook game log. Click on the first link, ESPN, and copy the last column into Minitab. Regular season only, and not the monthly totals.

We will perform a hypothesis test and a confidence interval to determine if Russell Westbrook has improved in 2016.

Hypothesis Test
1. H$_0$:
2. H$_A$:
3. Test statistic and its value:
4. *p*-value:
5. Your statistical conclusion:
6. Your interpretation:

Confidence Interval
7. 95% Confidence interval:
8. Your statistical conclusion:
9. Your interpretation:

Confirmation
10. Do the results of the confidence interval confirm the hypothesis test?

Appendix 1: Chi-Squared Distribution Table

The values in the body of the table are the critical values of chi squared.

degrees of freedom	Probability of a larger value of x^2 (p-value)									
	0.995	0.99	0.975	0.95	0.9	0.1	0.05	0.025	0.01	0.005
1	---	---	0.001	0.004	0.016	2.706	3.841	5.024	6.635	7.879
2	0.010	0.020	0.051	0.103	0.211	4.605	5.991	7.378	9.210	10.597
3	0.072	0.115	0.216	0.352	0.584	6.251	7.815	9.348	11.345	12.838
4	0.207	0.297	0.484	0.711	1.064	7.779	9.488	11.143	13.277	14.860
5	0.412	0.554	0.831	1.145	1.610	9.236	11.070	12.833	15.086	16.750
6	0.676	0.872	1.237	1.635	2.204	10.645	12.592	14.449	16.812	18.548
7	0.989	1.239	1.690	2.167	2.833	12.017	14.067	16.013	18.475	20.278
8	1.344	1.646	2.180	2.733	3.490	13.362	15.507	17.535	20.090	21.955
9	1.735	2.088	2.700	3.325	4.168	14.684	16.919	19.023	21.666	23.589
10	2.156	2.558	3.247	3.940	4.865	15.987	18.307	20.483	23.209	25.188
11	2.603	3.053	3.816	4.575	5.578	17.275	19.675	21.920	24.725	26.757
12	3.074	3.571	4.404	5.226	6.304	18.549	21.026	23.337	26.217	28.300
13	3.565	4.107	5.009	5.892	7.042	19.812	22.362	24.736	27.688	29.819
14	4.075	4.660	5.629	6.571	7.790	21.064	23.685	26.119	29.141	31.319
15	4.601	5.229	6.262	7.261	8.547	22.307	24.996	27.488	30.578	32.801
16	5.142	5.812	6.908	7.962	9.312	23.542	26.296	28.845	32.000	34.267
17	5.697	6.408	7.564	8.672	10.085	24.769	27.587	30.191	33.409	35.718
18	6.265	7.015	8.231	9.390	10.865	25.989	28.869	31.526	34.805	37.156
19	6.844	7.633	8.907	10.117	11.651	27.204	30.144	32.852	36.191	38.582
20	7.434	8.260	9.591	10.851	12.443	28.412	31.410	34.170	37.566	39.997
21	8.034	8.897	10.283	11.591	13.240	29.615	32.671	35.479	38.932	41.401
22	8.643	9.542	10.982	12.338	14.041	30.813	33.924	36.781	40.289	42.796
23	9.260	10.196	11.689	13.091	14.848	32.007	35.172	38.076	41.638	44.181
24	9.886	10.856	12.401	13.848	15.659	33.196	36.415	39.364	42.980	45.559
25	10.520	11.524	13.120	14.611	16.473	34.382	37.652	40.646	44.314	46.928
26	11.160	12.198	13.844	15.379	17.292	35.563	38.885	41.923	45.642	48.290
27	11.808	12.879	14.573	16.151	18.114	36.741	40.113	43.195	46.963	49.645
28	12.461	13.565	15.308	16.928	18.939	37.916	41.337	44.461	48.278	50.993
29	13.121	14.256	16.047	17.708	19.768	39.087	42.557	45.722	49.588	52.336
30	13.787	14.953	16.791	18.493	20.599	40.256	43.773	46.979	50.892	53.672
40	20.707	22.164	24.433	26.509	29.051	51.805	55.758	59.342	63.691	66.766
50	27.991	29.707	32.357	34.764	37.689	63.167	67.505	71.420	76.154	79.490
60	35.534	37.485	40.482	43.188	46.459	74.397	79.082	83.298	88.379	91.952
70	43.275	45.442	48.758	51.739	55.329	85.527	90.531	95.023	100.425	104.215
80	51.172	53.540	57.153	60.391	64.278	96.578	101.879	106.629	112.329	116.321
90	59.196	61.754	65.647	69.126	73.291	107.565	113.145	118.136	124.116	128.299
100	67.328	70.065	74.222	77.929	82.358	118.498	124.342	129.561	135.807	140.169